D0560857

DATE			

Food Systems and Agrarian Change

Edited by Frederick H. Buttel, Billie R. DeWalt,
and Per Pinstrup-Andersen

A complete list of titles in the series appears at the end of this book.

THE NEW ECONOMICS OF INDIA'S GREEN REVOLUTION

Income and Employment Diffusion in Uttar Pradesh

Rita Sharma
Thomas T. Poleman

Cornell University Press

ITHACA AND LONDON

First published 1993 by Cornell University Press.

International Standard Book Number 0-8014-2806-8
Library of Congress Catalog Card Number 92-56784
Printed in the United States of America
*Librarians: Library of Congress cataloging information
appears on the last page of the book.*

♾ The paper in this book meets the minimum requirements
of the American National Standard for Information Sciences—
Permanence of Paper for Printed Library Materials, ANSI Z39.48-1984.

To Vijai, Charlotte, and Lillian

Contents

Preface   xiii

Terms and Abbreviations   xvii

1 Technical Change and Rural Poverty   1

   Equity Issues: The Early Assessments 2
     Widening Regional Disparities 2
     Income Differentials 5
     Employment Aspects 8
     Agrarian Unrest 10

   Broader Implications of the Early Assessments 11
     Quantifying the Indian Case 12

   Gains from Green Revolution Technology 14
     Increased Food Production 14
     Making the Small Landholding Viable 16

   Emerging Trends: Second-Generation Effects 18
     Broadening the Regional Impact 18
     Reducing Income Disparities 19

   Organization of the Book 20

2 The Green Revolution in Uttar Pradesh: The Widening
   and Subsequent Narrowing of Regional Disparities   23
   Profile of Uttar Pradesh 23

Geographical Characteristics 24
Agrarian Structure 27

Historical Perspective on East-West Disparities 30
Causes of Uneven Development 32
Growth of Infrastructure 34
Is History Repeating Itself? 35

State Intervention to Correct Imbalances 36
Land Reform 36
Reorganizing Irrigation Priorities 37

Widening and Subsequent Narrowing of Regional
Disparities 38
Agricultural Performance: Macro Evidence 38
The Growing Importance of Rural Growth Centers 50
Development of Infrastructure 54

Income Differentials: Evidence from Micro Studies 55
Initial Indicators of Growing Inequality 56
Subsequent Signs of Improvement 57

Appendix 59

3   Income Diffusion over Time: Walidpur Village   63
Profile of Walidpur Village 66
Comparison with Meerut District and Western
Uttar Pradesh 67
Walidpur and Daurala Growth Center 68

The Benchmark and Present Studies 69
Data Collection 70

Impact of the Green Revolution 71
Changes in Agrarian Structure 72
Changes in Agricultural Performance 76

Income Diffusion over Time 81
Changes in the Level of Household Income 82
Changes in the Composition of Household Income 84
Changes in Income Distribution 85
Anticipating Future Income Distribution 86

Appendix 90

Photographs   97

4   Mechanisms of Income Diffusion: Meerut District   105
    Profile of Meerut District 105
       Demographic Characteristics 106
       Agricultural Performance 106
       Reasons for Selecting Meerut 108
    Mechanisms of Income Diffusion 110
       Off-Farm Diversification 110
       Dairying as a Source of Income Diffusion 115
       Cultivation of High-Value, Labor-intensive Crops 117
    Methodology and Data Collection 119
       Selection of Villages for Case Studies 119
       Estimation of Household Income 122

5   Off-Farm Diversification: Rampur Village   127
    Profile of Rampur Village 127
       Changes in Agrarian Structure 129
       Changes in Agricultural Performance 130
       Changes in Employment Patterns 130
    Increasing Importance of Off-Farm Activities 131
       Appeal of Off-Farm Employment 133
       Impact of Off-Farm Employment on Crop Cultivation 134
       Factors Influencing Off-Farm Employment 136
    Income Diffusion in Rampur 140
       Diversification of Household Income 140
       Implications for Income Distribution 142
    Household Economic Strategies 144
       Resources and Objectives 144
       Gaining Access to Capital 146
       Gaining Access to Skills and Education 147
    Case Studies 147
       Landless Households 148
       Marginal Households 157
    Appendix 168

6   Dairying as a Source of Income Diffusion: Izarpur Village   176
    Profile of Izarpur Village 176
       "Occupants of the Front Room" 177

Changes in Agrarian Structure 178
Changes in Employment Patterns 179
Changes in Agricultural Performance 180

The Appeal of Dairying to Small Producers 181
Participation of Small Producers 184
Factors Influencing Dairying Activities 187

Income Diffusion in Izarpur 190
Income from Dairying 190
Implications for Income Distribution 191

Household Strategies 193

Case Studies 194
Landless Households 194
Marginal Households 201

Appendix 205

7 Labor-intensive Cultivation of High-Value Crops:
   Jamalpur Village   211
Profile of Jamalpur Village 211
Changes in Agrarian Structure 213
Changes in Employment Patterns 214
Changes in Cropping Patterns 214

*Growing Importance of High-Value, Labor-intensive
Crops* 217
Appeal of High-Value, Labor-intensive Crops 217
Changes in Cropping Patterns 218
Obstacles to Adoption by Marginal Landholders 220

Income Diffusion in Jamalpur 224
Implications for Income Distribution 224

Household Strategies 226

Case Studies 228

Appendix 234

8 Speeding up Income and Employment Diffusion: Some
   Policy Implications   239
Summary of Findings 240
Regional-Level Comparison 240
Village-Level Studies 241

Household Case Studies 241
Is the Case of Meerut and Western Uttar Pradesh
   Applicable to Other Regions of India? 242
Hastening the Income Diffusion Process 243
Encouraging the Rural Off-Farm Sector 244
Encouraging Animal Husbandry 251
Improving Government Antipoverty Programs 253
Summing Up 254

Appendix   Decomposition of the Gini Coefficient by
   Sources of Income   256
References   259
Index   269

Preface

Twenty years ago D. K. Freebairn and I organized one of the first attempts to assess the social and economic consequences of the Green Revolution (Poleman and Freebairn 1973). The conclusion our group reached became for a time the accepted wisdom: that while the application of the scientific method to agriculture in the developing world held the potential for enormous increases in food production—far greater than any likely growth in population—the selectivity of its impact raised serious issues of equity. Because the high-yielding varieties that formed the core of the new technology required associated inputs of water, fertilizer, and pesticides in order to be effective, we reckoned that the lion's share of the benefits would go to favorably endowed regions, and within them to those farmers wealthy enough to afford the new inputs. The new technology therefore was likely to marginalize further the rural poor, and their efforts to make new lives for themselves in the cities would be frustrated by the inability of a capital-intensive industrialization process to absorb more than a fraction of them. To some this scenario meant that unless accompanied by institutional reform—or preferably preceded by it—the Green Revolution in the countryside could give rise to a Red one in town.

Time and further reflection suggested that these conclusions were somewhat hasty. Technical change in agriculture is almost invariably selective in its impact, and it does not follow that the impetus to growth it generates will not eventually diffuse into other sectors of the economy. A prospering agriculture requires more inputs, more processing, and more construction. On the consumption side, higher

xiii

farm incomes mean a growing market for a host of goods and services. The rich, to be sure, become richer, but the poor also gain. Irma Adelman calls this process, when dressed as a strategy for national economic development, "agricultural demand-led industrialization" (Adelman 1984, Adelman and Taylor 1990).

The present work considers technical change and its consequences in the Indian context, examining first the evolving assessments of the Green Revolution and then illustrating how the process of income diffusion works. Through an investigation of the changes that have occurred in Uttar Pradesh—India's largest state, its agricultural heartland, and one of the principal locales of the Green Revolution—we show that the impact of technical change has varied over time. Although regional differences widened initially, they subsequently narrowed as improved infrastructure made possible the diffusion of technology into hitherto bypassed areas. We also show that the second-generation effects of the Green Revolution—once dismissed as "trickle down"—have begun to be reflected in a host of new noncrop and off-farm employment opportunities. Participation of the landless and near-landless in these enterprises dispels the conclusion that technical change pauperizes the rural poor and points to an alternative path for employment and income diffusion, midway between the agricultural and urban-industrial sectors.

These are conclusions of far-reaching significance, ones that touch on the role governments in the developing countries should play now that the need to create more jobs has replaced increasing food production as the key challenge. A. K. Sen's (Dreze and Sen 1989) proposals to provide greater "entitlements" through more and larger welfare schemes have a simplistic appeal. But as Inderjit Singh (1990) noted in his excellent book on the rural poor in South Asia, hardly any official recognition has been given to noncrop and off-farm activities as sources of income for small farmers and the landless, let alone how they can be encouraged. In the last chapter we offer some suggestions, and hope that they will stimulate others to think on the matter.

This book is very much Rita Sharma's; my name as junior author appears as a matter of courtesy. The volume reflects almost three years of work on her part, eighteen months of which were spent in the field, and is the culmination of six years of graduate study at Cornell University. Dr. Sharma is currently Commissioner, Rural Development for the Government of Uttar Pradesh, and has for most of the past twenty years been involved with rural and agricultural development in Uttar Pradesh. In 1984 she was given leave from her position as Special Secretary in the Department of Agriculture and awarded a Hubert Humphrey Fellowship for study in the United States. She was subsequently awarded a Sage Fellowship by Cornell University and assistantships by the Department of Agricultural Economics, supplemented by fellowships from the Institute for the Study of World Politics, the P.E.O. Sisterhood, and the American Association of University Women. Her fieldwork was made possible by a grant from the Ford Foundation. To all we are indebted.

Many people contributed to the book and to the study on which it is based. Special thanks go to Sanjay Agarwal and D. S. S. Yadav in Meerut. Joseph Baldwin and Lillian Thomas drew the maps and charts, and Lillian Thomas also prepared innumerable revisions of the manuscript. For their comments and suggestions we are grateful to Randolph Barker, Billie DeWalt, D. K. Freebairn, Davydd Greenwood, Keith Griffin, Bruce Johnston, Tom Kessinger, R. J. McNeil, K. L. Robinson, J. S. Sarma, P. V. Sukhatme, and David Thurston.

Although their names are made up, Rampur, Sitapur, Izarpur, and Jamalpur are real places. Since their locations are accurately described in the text and maps, a minimum of sleuthing could reveal their actual names. We ask readers not to do this for it could cause embarrassment to those who greeted Dr. Sharma and her assistants with friendliness and unflagging cheer.

T. T. P.

Ithaca, New York

Terms and Abbreviations

Abadi	Residential area of a village
Agola	Sugarcane tops used as animal feed
Arhar	A pulse, also known as redgram or pigeon pea
Artia	Commodity broker, commission agent
Atta-chakki	A motor-driven grain mill
Bandobast	Land revenue settlement
BDO	Block Development Officer
Berseem	A clover planted as fodder during the *rabi* season
Bhaiachara	Coparcenary tenurial form in which the revenue demand is apportioned among village proprietors on some principle other than ancestral shares, usually on the basis of cultivation
Bhisti	Water carrier
Bhumidhar	A tenant who owns land in his or her own right and sells it if he or she wishes; under the Zamindari Abolition Act, tenants of the *zamindars* became *sirdars* (tenants of the state), but they could convert to *bhumidhari* status on payment of ten times land revenue
Bithaura	Stacks of dried dung cakes
Block	An administrative subdivision of a district, headed by a BDO
Buggi	A cart with tires instead of wooden wheels, usually pulled by a male buffalo
Charpoy	Rope bed
Cheri	Sorghum fodder
Chulha	Clay stove
Dhobi	Washerman

xvii

Doab	Area between rivers running roughly parallel
DRDA	District Rural Development Agency
Dudhiya	Milk vendor who purchases from the producer and sells to the dairy
Dunlop cart	See *Buggi*
Foodgrains	Includes cereals and pulses
Ghee	Clarified butter
Gur	Cakes of unrefined sugar made by boiling sugarcane juice
HYV	High-yielding varieties
IRDP	Integrated Rural Development Program
Jajmani	Patron-client relationship
Jat	A farming caste
Jawan	Soldier in the Indian army
Jowar	Sorghum
Kharif	Rainy season: sowing in June–July, harvest in October–November
Khasra	Record of cropping pattern in the village
Khatauni	Record of landownership in the village
Kolhu	A motor-driven sugarcane-crushing unit
Kumhar	Potter
Kutcha	Roughly made, the opposite of *pucca*
Lahi	Early mustard
Makka	Maize
Mandi	Regular market yard
Maurusi	Occupancy status
Mawa	Solid residue from evaporating milk, used for preparing sweets and desserts
Mung	A pulse, also known as greengram
Painth	Periodic market for a group of villages
Panchayati	A three-tier system of local self-government at the *Raj* village, block, and district levels
Parchoon	Miscellaneous items such as cigarettes, matches, tea, salt, sugar, spices, and cooking oil
Pradhan	Elected headman of the village
Pucca	Of a house or building constructed of brick, not mud
Rabi	Winter cropping season: sowing in November–December, harvest in April–May
Rupee	The basic unit of currency in India; at the time of the study, 1988–89, one U.S. dollar was worth about fifteen rupees
Scheduled caste	Groups mentioned in the Constitution of India as deserving special favor in view of their former oppressed position
Sirdar	Tenant of the state (see *Bhumidhar*)

Talukdar	Large landowner
Teli	Oil pressers
Tempo-taxi	Three-wheeled vehicles, seating eight to ten people, used as a mode of transport
Tons	Unless otherwise noted, all tons are metric tons
TRYSEM	Training Rural Youth for Self-Employment
Urd	A pulse, also known as blackgram
Zaid	Dry summer season after the harvest of the *rabi* crop
Zamindar	A hereditary agent who held land, paid revenue to the government of the Moghuls and later to the British, and acted as an intermediary between the authorities and the cultivators of the land

The New Economics of India's Green Revolution

1

Technical Change and Rural Poverty

Following the introduction to India of high-yielding varieties of wheat and rice in the late 1960s and early 1970s, there arose a widespread belief that modern technology was very much a double-edged sword. Studies of the initial impact of the Green Revolution acknowledged the positive contribution of modern varieties in increasing foodgrain production and laying to rest the Malthusian nightmare. But the consensus seemed also to be that the new technology had caused resource-rich regions to gain at the expense of ones less well endowed and big, wealthy farmers to benefit more than the landless and near-landless. These early studies suggested that this process of agricultural change would promote polarization of the rural society and marginalization of the rural poor. Some social scientists saw the new technology sharpening social and political tensions.

The purpose of this book is to challenge that assessment and to suggest that whereas the initial impact of technical change indeed supported a pessimistic point of view, subsequent developments do not. Through an examination of the changes that have occurred in the state of Uttar Pradesh, particularly the western region, one of the principal locales of the Indian Green Revolution, we demonstrate that the interregional and interpersonal impact of the new technology has differed over time. Our findings suggest that while regional disparities widened in the initial phase—from the mid-1960s to the mid-1970s—subsequent years have witnessed a narrowing of differences as technology has diffused into some of the hitherto bypassed regions in the wake of improved infrastructure. Moreover, the second-generation ef-

1

fects of the Green Revolution have begun to be reflected in the variety of new economic opportunities created as a result of agricultural growth. Increased participation of the rural poor in these noncrop activities has led to an improvement in their income level and to a reduction in poverty.

Equity Issues: The Early Assessments

The initial perceptions of the Green Revolution in India emphasized the selective nature of technical change. Attention was drawn to the fact that the high-yielding varieties (HYVs) required a complementary package of inputs, including fertilizer, controlled irrigation, and chemicals for plant protection. Ecological and institutional constraints in obtaining these purchased inputs limited the spread of technology to regions with well-developed infrastructure and restricted the number of producer beneficiaries to those who could afford the package.

Widening Regional Disparities

Most observers agree that the new technology increased regional differentiation in India. Areas endowed with favorable natural resources and well-developed physical and institutional infrastructure were the first to experience the spectacular gains in cereal output resulting from the "miracle seeds." In regions lacking these facilities, adoption of the new technology was tardy and agricultural production stagnated (Bhalla and Alagh 1979, Bhalla and Tyagi 1989a, Easter et al. 1977, Frankel 1971, Krishnaji 1975, Ladejinsky 1969b, Prahladachar 1983, Staub and Blase 1974).

In its natural resources India is very unevenly endowed. Of the approximately 150 million hectares of land cultivated every year, about 100 million hectares have a mean rainfall ranging from about 35 to 140 centimeters. Most of the country can be classified as semiarid. A substantial area in the central and southern regions is arid and infertile; 50 centimeters of rain or less fall in just one or two months.

In sharp contrast is the corner of the country long recognized as one of the world's great breadbaskets: the northwestern Indo-Gangetic plain, made up of the states of Punjab, Haryana, and western Uttar Pradesh, all characterized by alluvial soils, abundant surface and groundwater, and medium rainfall. These natural advantages have been reinforced through irrigation, rural electrification, and consolidation of landholdings. The introduction in this region of canal irrigation in the mid-nineteenth century changed the orientation of farming from subsistence to commercial. The area possessed all the preconditions in the mid-1960s for rapidly exploiting the HYV package (Stokes 1978, Stone 1984).

G. S. Bhalla and D. S. Tyagi (1989a, 1989b)recently made a detailed analysis of the spatial disparities resulting from the unequal adoption of technology. They divided the post–Green Revolution period into an initial phase extending from the mid-1960s to roughly the mid-1970s and a later period running from the mid-1970s to the mid-1980s. By comparing agricultural growth rates for major crops in the two time periods, Bhalla and Tyagi found that the gap between the northwestern region and the rest of the country widened rapidly in the initial phase.

Figure 1.1 indicates changes in the share of agricultural production of major crops, by region, in the post–Green Revolution period. The northwestern region increased its share in total output from about one-fourth in the mid-1960s to more than one-third in the early 1980s, even though in terms of net cultivated area this region accounts for less than one-fifth of India's total. The central region, largest in terms of cultivated area, accounts for slightly less than half the net sown area yet contributes only a quarter of production. The southern and eastern regions show a persistent declining trend in output share.

A comparison of growth rates of agricultural output by state further highlights the variation in agricultural performance. The northwestern region registered an annual growth in agricultural output of about 5 percent during the first phase. The performance of Punjab and Haryana, with growth rates above 8 and 6 percent per annum, respectively, contributed largely to this impressive growth.

The central region, by contrast, had a growth rate of less than 1 percent per annum. Agriculture in this arid to semiarid region is primarily rainfed, as it is in the southern region, which experienced a growth rate of less than 1.5 percent. The eastern region, consisting of Bihar, West Bengal, Orissa, and Assam, often referred to as the "Achilles' heel" of Indian agriculture, seems to have gone from bad to worse. The 1.4 percent growth rate experienced in the initial phase of the Green Revolution declined to about 0.5 percent in the subsequent period. Although well endowed in terms of rainfall and alluvial soils, the eastern region is characterized by a lack of infrastructure, a high

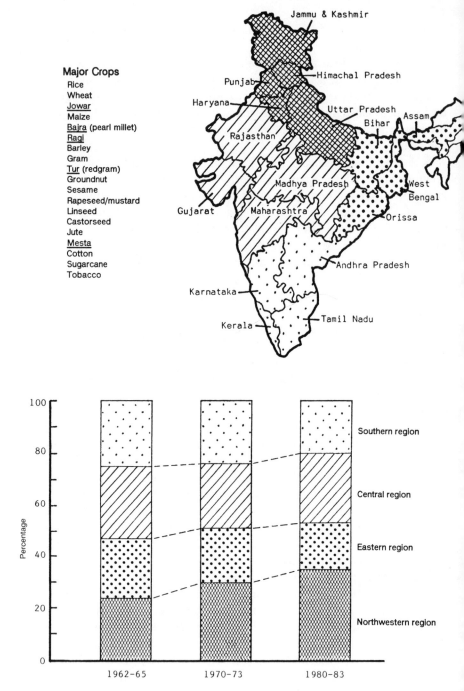

Major Crops
Rice
Wheat
Jowar
Maize
Bajra (pearl millet)
Ragi
Barley
Gram
Tur (redgram)
Groundnut
Sesame
Rapeseed/mustard
Linseed
Castorseed
Jute
Mesta
Cotton
Sugarcane
Tobacco

Source: Compiled from data in G. S. Bhalla and D. S. Tyagi, "Spatial Pattern of Agricultural Development India," *Economic and Political Weekly* 15, no. 25 (1989): A46–A56.

Figure 1.1. Share of agricultural output for major crops by region and state in India, 1962–196. 1970–1973, and 1980–1983

man-land ratio, and annual flooding. Modern technology has been unable to make a significant impact there.

These widening regional disparities reflect differences in the extent and timing of HYV wheat and rice adoption. Punjab, Haryana, and western Uttar Pradesh, where the technology was adopted in the mid-1960s, are now experiencing the second-generation effects of technical change—the diversification of their rural economies. The spread of the HYV technology began almost a decade later in the eastern region and along the coastal strips of peninsular India, and its potential is yet to be fully exploited. In the arid and semiarid tracts of the central and southern regions, the HYV wheat and rice technology has scarcely made any headway at all.

Income Differentials

Early critics of the Indian Green Revolution argued that the selective nature of the new technology did not just confine its benefits to a few favored parts of the country. Even within well-endowed regions, they noted, new production techniques were primarily captured by and benefited the rural elites. The less privileged majority—near-landless cultivators, tenants, and landless laborers—were bypassed (Dasgupta 1977; Frankel 1971; Griffin 1989; Ladejinsky 1969a, 1969b; Pearse 1980).

Biplab Dasgupta (1977:372) concluded that among the unfavorable consequences of technical change were "proletarianization of the peasantry and a consequent increase in the number and proportion of landless households, growing concentration of land and assets in fewer hands, and widening disparities between the rich and poor households." Commenting on the situation faced by the less privileged rural groups, A. C. Pearse (1980:181) observed: "A misleading scale neutrality was claimed from the new technology on the basis of the divisibility of seeds and chemicals, its main components. In fact, the socioeconomic magnitude of the cultivator is of utmost importance for his economic success, where he must compete with well capitalized large farmers."

Countering this argument, others (Barker et al. 1985, Dantwala 1986b, Hayami and Kikuchi 1981, Pinstrup-Andersen and Hazell 1985) contended that the fault lay not so much in the nature of technology, which was divisible and scale neutral, as in structural rigidities in social and agrarian institutions that resulted in uneven distribution of gains between different social groups. Institutional constraints prevented equal access to inputs and enabled the richer minority of the rural population to monopolize the technology and its gains. Nonimplementation of structural reforms, population pressures, labor-saving mechanization, and a constricting land base were held responsible for the growing incidence of poverty rather than the Green Revolution technology itself.

Keith Griffin (1989:147) rebutted this line of reasoning by noting that "the purpose of the Green Revolution strategy was precisely to circumvent the need for institutional change. Technical progress was regarded as an alternative to land reforms and institutional transformation—the green revolution was to be a substitute for the red—and it is misleading twenty years later to claim that there was nothing wrong with the original strategy and that the fault lies entirely with inappropriate institutions and policies."

The crucial institutional constraint is the unequal distribution of land. Demographic pressures are leading to a rapidly declining land-man ratio. The average size of an operational holding declined from 2.5 hectares in the mid-1950s to about 1.8 hectares in 1980–81 (Dantwala 1986a).

The total number of operational landholdings in the early 1980s was around 90 million (India, Ministry of Agriculture 1987). Landholdings are officially categorized into five groups: *marginal* (below 1 ha), *small* (1–2 ha), *semi-medium* (2–4 ha), *medium* (4–10 ha), and *large* (above 10 ha). The changing percentage distribution of holdings and area by category is shown in Figure 1.2. Marginal holdings, with an average area of less than 0.5 hectare, accounted for more than half of all holdings in 1980–81, yet operated just over one-tenth of the total area. Even more noteworthy is the rapid subdivision of land into marginal holdings, which rose by nearly 15 million in the space of a decade, an increase of over 40 percent. At the present rate more than two-thirds of all holdings will be marginal by the turn of the century. At the other end of the spectrum, large holdings represent about 2 percent of holdings but account for almost one-fourth of the operated area.

Land remains the single most important determinant of agrarian relationships. Access to other rural factors of production is largely determined by the position of the individual in the land hierarchy. Most assets other than land also belong to those who own land (Reserve Bank of India 1976). Indeed, the distribution of other assets is even more skewed than that of landownership. Overall, nearly three-fourths of all rural assets are owned by the top quarter of households,

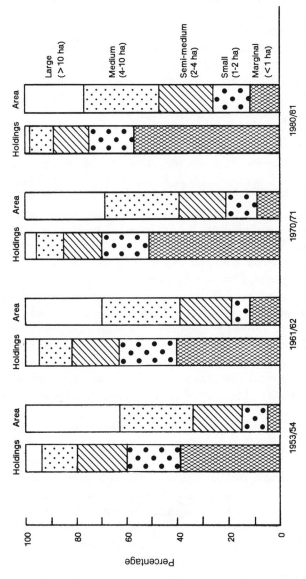

Large
(>10 ha)

Medium
(4-10 ha)

Semi-medium
(2-4 ha)

Small
(1-2 ha)

Marginal
(<1 ha)

Area Holdings   Area Holdings   Area Holdings   Area Holdings
1980/81   1970/71   1961/62   1953/54

Percentage

Sources: India, Ministry of Agriculture, Directorate of Economics and Statistics, *Indian Agriculture in Brief* (Delhi, 1986); India, Cabinet Secretariat, *The National Sample Survey* (Delhi: Controller of Publications, 1953–54, 1961–62).

Figure 1.2. Percentage distribution of operational holdings and area in India, by size group, 1953–1954 to 1980–1981

7

while the bottom 25 percent owns less than 2 percent (Divatia 1976). Ownership of land usually coincides with control of local institutions, and those without land are severely handicapped. Access to credit is denied them for lack of collateral. Their low socioeconomic status, defined by limited landownership, prevents fair access to information and extension facilities. Not surprisingly, big farmers with a substantial land base, capital resources, access to irrigation, fertilizer, credit, extension services, marketing, and information, and political clout were the first beneficiaries of the Green Revolution (Lockwood et al. 1971, Sen 1985). As early adopters, they reaped the major gains from innovation (Quizón and Binswanger 1986).

Many studies were conducted in the late 1960s and the 1970s to evaluate the impact of technical change on income distribution between various categories of farms (Bardhan 1974, Bhalla and Chadha 1981, Clift 1977, Garg et al. 1972, Junankar 1975, Saini 1976b, Srivastava et al. 1971). Although not unanimous in their conclusions, most tended to support the notion of gains for the big farmers and a worsening income distribution. Profits from the early adoption of the new technology led to farm mechanization and the investment of even more capital. Coupled with the fact that HYVs had enhanced the value of land, there was a tendency to enlarge farm size through eviction of tenants, purchase of land, and in some cases reverse tenancy, in which small owners rent land to larger landowners (Frankel 1971; Ladejinsky 1969a, 1969b; Singh 1989). As Wolf Ladejinsky (1969b: 160) put it: "The sharecroppers are, if anything, worse off now than before because as ownership of improved land is prized very highly there is mounting determination among owners not to permit the tenants to share in the rights of the land they cultivate. Their preference is to be rid of them."

Employment Aspects

An accurate picture of the impact of technological change on demand for labor would require a much larger canvas than has generally been used in field studies. The direct impact or first-round effects of HYV technology is usually estimated with farm-level studies, but this can lead to conflicting conclusions. Additional man-days of employment created through multiple cropping and an enhanced volume of production (Chinnappa 1977, Herdt 1980) commonly result in increased wage rates for agricultural labor (Lal 1976, Rao 1975). On the other hand, when labor has been displaced by mechanization, especially tractors (Billings and Singh 1970), the effect can be the opposite, and the decline in real wage rates for farm workers in the post–Green Revolution period has caused more than a few commentators to conclude that the overall impact of technology on labor demand is negative (Bhalla 1979).

These studies generally limited the measurement of demand for labor to agricultural operations. The impact of the Green Revolution, however, has traveled far beyond the farm, and the second-generation effects of technical change have brought increased demand for both agricultural and nonagricultural goods and services, resulting in greater employment opportunities in postharvest operations, storage, milling, marketing, and transport, and extending the arena of influence of the agricultural technology from the farm to market yards, rural towns, warehouses, and railway stations. The new technology has given rise to a host of activities to support and sustain the modern production process. However, such widespread effects are hard to quantify, and consequently hardly any studies focus on the second-generation impact of agricultural technology. These effects were almost completely ignored in the early assessments of the Green Revolution.

At an aggregate state level, however, there is evidence to suggest that agricultural growth may have improved employment prospects. States such as Punjab, Haryana, and Uttar Pradesh, which registered impressive agricultural performance in the early 1970s, also recorded the lowest unemployment rates during this period. Seasonal variations in employment were also the lowest in these states, indicating the effects of increased multiple cropping (Grawe, Krishnamurti, and Bassh-Dwomo 1979).

Despite the undisputed gains in foodgrain production and the improvement in employment opportunities in some states, the Green Revolution did not make a major dent in rural poverty in the country as a whole. The proportion of the rural population with incomes below the official poverty line[1] ranged between 40 and 55 percent from the mid-1950s to the late 1970s. Even in the post–Green Revolution period there have been years when as much as 50 percent of the population had incomes below the poverty line (Ahluwalia 1986). The average incidence of officially recognized rural poverty in the pre– and post–Green Revolution era declined by only 4.5 percentage points (Griffin 1989).

---

1. The official poverty line is defined in Chapter 3.

In defense of the Green Revolution technology, M. L. Dantwala (1986b:120) observed that "there is no evidence of either reduced absorption of hired labor in agricultural production or reduced real wage under the 'new technology.' The more likely explanation for the increased incidence of poverty is that prosperity led to in-migration from backward regions. If poverty has increased in spite of higher food production, nonagricultural factors such as inflation, demographic pressures and failure to diversify the rural economy and reduce the dependence of the growing labor force on farm land should be held responsible."

Agrarian Unrest

For centuries village communities have been held together through institutions of mutual aid and patron-client relationships, which ensured a minimal subsistence for all and contributed to social stability. According to the writings of Marx (1925) and Lenin (1960), the introduction of capitalist production techniques leads to a breakdown of this stability. Modern capital-intensive agricultural technology causes social obligations to be reevaluated in commercial terms. Big, wealthy farmers, operating on motives of profit maximization, find it profitable to evict their tenants, buy out small farms, and substitute formal contracts for customary arrangements. The less-privileged majority is forced to earn an increasing proportion of its income from the sale of labor and finds its position deteriorating.

We summarize the argument in this simplified form because critics of the Green Revolution have drawn support from this line of reasoning to assert that the selective nature of modern agricultural technology has exacerbated the situation in favor of the big farmers, accelerating the polarization process and leading to the marginalization of the landless and near-landless (Byres 1981, Frankel 1971, Griffin 1974, Patnaik 1987, Pearse 1980, Sanyal 1988, Sharma 1973). Some have extended the argument to predict political instability and agrarian unrest as a result of the Green Revolution. Francine Frankel's writings have been among the most influential in this regard. She noted that "the introduction of modern technology under the intensive areas and the high-yielding varieties programs has not only quickened the process of economic polarization in the rural areas, but it has also contributed to increasing social antagonism between landlords and tenants, and landowners and laborers" (Frankel 1971:197).

The more radical spokesmen for this line of reasoning warned that the Green Revolution carried within it the seeds of a Red one.

Broader Implications of the Early Assessments

Early assessments of the Green Revolution implied an apparently intractable situation for employment and rural poverty. A historical parallel can be drawn with the agricultural revolution that took place in Europe and North America in the nineteenth century, when adoption of new agricultural techniques displayed a similar selectivity in impact. Productivity increases allowed fewer farmers to produce increasing amounts of food, thereby rendering surplus a large number of rural workers. Dispossessed tenants, bypassed laborers, and out-of-work artisans migrated to cities in search of employment and were absorbed in manufacturing, construction, trade, and services. The transition from a predominantly rural society to an urban one defines the classical model of development. However, a number of differences make it difficult for the Western experience to be replicated today.

The sheer magnitude of the current labor force confounds the absorptive capability of urban areas. A century and a half ago, when the presently developed countries were experiencing the transition, the demographic pressures were nowhere as intense as they are today. Projections made in 1971 by the International Labor Office estimated that the labor force in developing countries would double from about one to two billion people between the years 1970 and 2000. The billion new jobs required in the span of three decades are almost twice the total number existing in the developed countries at present and provide an indication of the absolute magnitude of the labor force growth (Poleman 1989). It is estimated that in India alone about 150 million new jobs would be required between 1980 and 2000 for new entrants into the work force (Sundaram 1984).

Furthermore, the cities in the developing world are already over-populated and cannot cope with large-scale migration from rural areas. In India in the decade of the 1970s, urban population grew at the rate of around 4 percent per annum—more than double the rural population growth. This figure conceals even more impressive increases in the growth of major cities. By the year 2000 Calcutta and Bombay are each expected to have more than 15 million inhabitants (Finance and Development 1989).

Even if the infrastructure could survive the onslaught of a large number of migrants, there would still remain the problem of gainful employment. A hundred and fifty years ago, the agricultural revolution in the West coincided with the onset of the Industrial Revolution. The labor-intensive nature of manufacturing, construction, trade, and commerce generated a ready demand for labor, and most who left the land found jobs (Poleman 1989). The modern industrial sector in today's developing countries is more capital- than labor-intensive and has not fulfilled its traditional role as absorber of redundant manpower from agriculture.

Lastly, populations in the developing countries today have a limited opportunity for emigration. This is in contrast with the situation a hundred and fifty years ago, when the Americas and Oceania provided a convenient outlet for the growing population of Europe. An estimated 35 million people migrated from Europe in the nineteenth century. No such prospects exist today. The solution to problems of rural unemployment must be found within the confines of the country, preferably in the rural areas.

Quantifying the Indian Case

Two simple analyses help to highlight the magnitude of surplus labor in agriculture and the employment dilemma confronting rural India. The past two and a half decades of development witnessed a decline of nearly 20 percent in the contribution of agriculture to the country's GDP. However, the corresponding reduction in labor force in the agricultural sector was a mere 3 percent (Fig. 1.3). In the same period, while industry enhanced its share in the GDP by seven percentage points, its share of labor absorption increased by a paltry 2 percent. Over the past three decades more than two-thirds of the total work force has continued to draw sustenance from the land. Industry has been incapable of providing employment to more than one-seventh of the labor force.

In the early years of planned development in India the basic tenet was to give priority to large-scale capital goods industries and the latest capital-intensive techniques to modernize plants and equipment. The concept of "intermediate technology," by which is meant labor-intensive technology, has not yet made any significant impact on the employment problem. It is safe to conclude, therefore, that the prospect of redundant rural labor finding gainful employment in the ur-

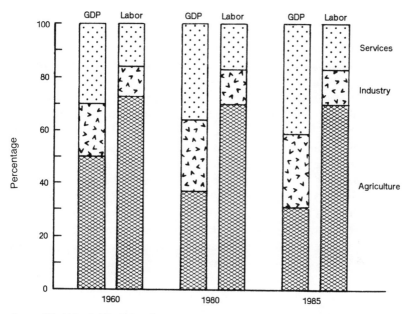

Source: World Bank, *World Development Report* (Washington, D.C.: World Bank, various years).

Figure 1.3. Percentage distribution of gross domestic product and labor force in agriculture, industry, and services in India, 1960, 1980, and 1985

ban-industrial sector appears bleak, at least within the next two decades.

The rural poor, especially the landless and the near-landless, do not seem to have fared much better in the agricultural sector. Consider the evidence regarding landholdings. In 1980–81 the total number of landholdings in the country was around 90 million. Of these, about 50 million, or more than half, were marginal (<1 ha) and 16 million were in the small (1–2 ha) category. Studies indicate that landholdings falling into the small category can become economically viable with irrigation and the use of modern varieties (Bhalla 1979, Dantwala 1973). However, the same optimistic prospect does not hold for marginal holdings. The average size of a marginal holding is less than 0.5 hectare. Crop cultivation on such very small plots of land, despite the use of a modern package of practices, does not provide an acceptable subsistence to farming households (Bhalla and Chadha 1983, Chatterjee et al. 1982). Farmers operating marginal landholdings must therefore look for alternate sources to supplement family income. There were almost 50 million such households in India in the early 1980s.[2] In addition to households operating marginal holdings, there were in the early 1980s around 10 million landless households (India, Ministry of Labour 1976) unable to earn a satisfactory subsistence solely from agricultural labor.

We thus conclude that at least 60 million rural households in the early 1980s were unable to adequately support themselves from crop husbandry and agricultural labor alone. The total number of rural households at this time was about 105 million[3] (India, Office of the Registrar General, *Census of India*, General Population Tables). This implies that about 60 percent of the 105 million rural households found it difficult to subsist on agriculture alone and had to look for alternate sources to supplement their income. Mahendra Dev has estimated that about 80 million workers will have to move out of agriculture by the year 2000 if the man-land ratio is to be maintained at the level of the year 1977–78 (Dev 1988).

Gains from Green Revolution Technology

Despite criticism of the Green Revolution on grounds of equity, the ability of modern varieties to increase total foodgrain production has never been questioned. Moreover, the technology has been responsible for making many small farms economically viable, where previously only big and medium farms had that capability.

Increased Food Production

Green Revolution technology has been the major instrument behind the impressive gains in foodgrain output in India. Production has doubled since the mid-1960s, when the HYV seeds were first introduced (India, Ministry of Agriculture, *Agricultural Situation in India*). With the ability of foodgrain production to keep pace with population expansion, fears of the Malthusian prophesy were allayed. In addition, the danger of severe food scarcities became a thing of the past, as did the country's dependence on imports. Unlike the droughts of the mid-1960s, severe droughts in 1979–80 and 1987–88 were tided over largely from reserve domestic stocks (Rao et al. 1988).

2. Assuming a one-to-one correspondence between landholdings and households.
3. Assuming five members per rural household.

Figure 1.4 presents the growth in production of wheat, rice, and total foodgrains and total population. The impact of the HYV technology on wheat production has been spectacular. With a growth rate of nearly 6 percent per annum, wheat output quadrupled in the post– Green Revolution period. Similar gains were not visible for foodgrains as a whole primarily because other crops have been unable to keep pace with wheat. Rice, which accounts for more than half the foodgrain output, registered only a 2.4 percent growth rate, while

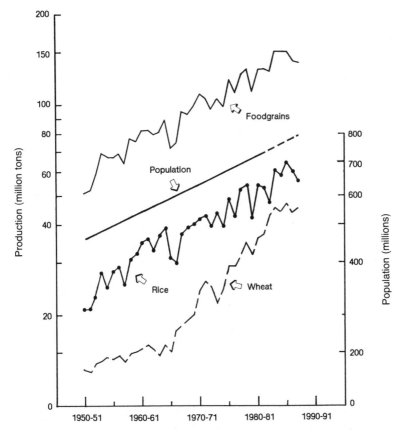

Sources: India, Ministry of Agriculture, Directorate of Economics and Statistics, *Area and Production of Principal Crops in India* (Delhi, various years); India, Ministry of Agriculture, Directorate of Economics and Statistics, *Indian Agriculture in Brief* (Delhi, 1986).

Figure 1.4. Production of foodgrains, rice, and wheat, and growth of population in India, 1950–1951 through 1987–1988 (vertical logarithmic scale)

pulses exhibited a negative trend compared with their pre–Green Revolution performance. Coarse cereals remained unchanged.

Despite wide variations in the performance of individual crops and regions, total foodgrain production maintained a growth of about 2.7 percent per annum, which just kept ahead of population growth (Bhalla and Tyagi 1989a). Not surprisingly, there have been no dramatic improvements in per capita food supplies, although a smaller proportion of the population than in the early 1950s has an inadequate diet. The impact of technical change is highlighted by the fact that whereas production gains were achieved largely through area expansion before the mid-1960s, in the subsequent period yield increases have been the major factor. This is illustrated in Figure 1.5, which compares production, area, and yield of foodgrains. In the pre–Green Revolution period, area expansion was responsible for nearly half the increase in production; in the post–Green Revolution phase, the contribution of area was reduced to one-tenth.

Making the Small Landholding Viable

One of the positive outcomes of Green Revolution technology has been to reduce the viability threshold of landholdings (Bhalla and Chadha 1981, Dantwala 1973). Studies show that whereas at least 4 to 6 hectares of unirrigated land was once required to sustain a household through crop income alone, a farmer cultivating 1 hectare of irrigated land can now expect to earn a satisfactory subsistence from his land. This has important implications for employment. In the early 1980s around 16 million landholdings were in the small category (1–2 ha), of which about half received irrigation (India, Ministry of Agriculture 1987). The HYV technology has thus made it possible for about half the households in the small farm category to support themselves through crop cultivation and holds out the same prospect for small farms for which irrigation potential exists but has not yet been exploited. The same, however, cannot be expected from marginal landholdings (<1 ha). Alternate means will have to be found, where they do not already exist, to supplement the crop income of the marginal farmer. Even with the adoption of seed-fertilizer technology, "very small or tiny holdings do not become viable" (Bhalla and Chadha 1983:153).

17

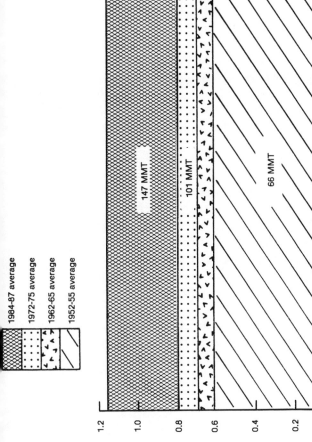

Source: India, Ministry of Agriculture, Directorate of Economics and Statistics, *Area and Production of Principal Crops in India* (Delhi, various years).
Note: MMT = million metric tons.

Figure 1.5. Production, area, and yield of foodgrains in India, 1952–1955, 1962–1965, 1972–1975, and 1984–1987 averages

Emerging Trends: Second-Generation Effects

The stimulus to local expansion and diversification of India's rural economy is closely related to the multiplier effects emanating from the growth generated by the new agricultural technology. The second-generation effects of technical change are manifest in the enhanced economic activity created to meet growing consumption demand, the result of higher rural incomes. These effects have given rise to a wide range of noncrop and off-farm employment opportunities. Especially noteworthy is that the landless and near-landless are increasingly able to participate in these enterprises. These activities do not require migration into cities; more and more they are located in rural growth centers that have developed around the existing nuclei of market towns, factory townships, and decentralized government offices.

Attention was first drawn to these second-generation effects in the early 1970s. The most prolific on the subject has been John W. Mellor, who emphasized linkages between the agricultural and nonagricultural sectors (Mellor 1976, 1986; Mellor and Lele 1973), especially consumption linkages resulting from enhanced rural incomes that have the potential for rural employment generation. He observed that "the strength of the growth linkage multipliers and their concentration on labor intensive goods and services produced within rural areas for local household consumption suggests that agricultural growth has the potential to significantly enhance rural nonfarm employment thereby broadening the participation of the poor in the benefits of growth and generating a greater market for agricultural output" (Mellor 1983:6). R. T. Shand (1983) also emphasized the importance of agricultural growth as a first step toward higher rural incomes.

The second-generation effects escaped the attention of most early commentators, whose assessments were based solely on accounts of farm operations and hence captured only the agricultural impacts of the Green Revolution. Furthermore, many of the second-generation effects were only partially visible at the time and thus escaped recognition.

Broadening the Regional Impact

Although the initial assessments noted that the Green Revolution was concentrated in the states of Punjab, Haryana, and western Uttar Pradesh, studies in the 1980s recognized that the HYVs of wheat and rice had extended into other regions, notably eastern Uttar Pradesh and coastal districts of Andhra Pradesh (Alagh 1982, Bhalla and Tyagi 1989a, Reserve Bank of India 1984).

Bhalla and Tyagi's (1989a) district-level study covering the mid-1960s through the mid-1980s demonstrates that there was a favorable development in the spread of technology in the latter part of the period. Many districts in Andhra Pradesh and eastern Uttar Pradesh that had previously been bypassed improved their productivity levels significantly in the 1980s.

Gilbert Etienne's (1988) study of Varanasi district (eastern Uttar Pradesh) also provides an indication of the regional diffusion of HYV technology. His time comparison of the agricultural performance of this district between 1967, 1978, and 1985 shows that Varanasi made impressive progress in the second period.

Reducing Income Disparities

The initial impression that the Green Revolution was exclusively a "big farmer phenomenon" contrasts with the picture emerging from later surveys and studies, which point to the universal adoption of HYV technology irrespective of farm size (Chinnappa 1977; Dasgupta 1977; Mandal and Ghosh 1976; Prahladachar 1983; India, Planning Commission 1976). Indeed, in the mid-1980s about 85 percent of wheat and more than 55 percent of the rice acreage in India was sown to high-yielding varieties (India, Ministry of Finance 1989). Once they were given the necessary infrastructural support, small farmers exhibited a remarkable tendency to catch up with the big farmers (Chadha 1979, Vyas 1979). Nowhere was this more pronounced than in the states of Punjab and Haryana, where government policies had a favorable "rural bias" (Dantwala 1986a, Westley 1986).

Few studies, however, have attempted to measure the impact of agricultural change and new employment opportunities on the diffusion of income in rural India.[4]

This book is intended as a first step toward correcting this.

4. Etienne (1988) compared district-level data on agricultural performance and wage rates in 1967, 1978, and 1985 for Bulandshahr (western Uttar Pradesh), Guntur (Andhra Pradesh), and Thanjavur (Tamil Nadu) and concluded that these districts, which experienced sustained agricultural growth, are now showing signs of diversification into off-farm activities. G. K. Chadha (1983), in his study of Punjab, indicated that a considerable proportion of industrial activity is related to agriculture. Consequently, such off-farm activity is located in small and medium-sized towns widely dispersed in the state. This enables rural

Organization of the Book

In the next chapter we use the state of Uttar Pradesh (U.P.) to demonstrate that the regional impact of technology has tended to vary over time. Uttar Pradesh is a natural laboratory because it encompasses a diverse range of geographical and historical experience. Wide variations are apparent in the development patterns of different regions in the state. Using macro data, we highlight the trend in regional disparities between western and eastern U.P. during the two post–Green Revolution periods. While the western region advanced more rapidly in the initial phase (mid-1960s to mid-1970s), several districts of eastern U.P. appear to have been catching up in terms of agricultural productivity and growth rates in the later period (mid-1970s to mid or late 1980s). The spread of infrastructure, primarily irrigation, has been responsible for the diffusion of technology in the east. The experience of Uttar Pradesh provides grounds for optimism, especially for regions with unexploited irrigation potential.

The remainder of the book focuses on Meerut District in western U.P., which presents an ideal site for observing the diversification of the rural economy. Meerut was one of the districts where the HYV technology had its earliest and most forceful impact in terms of increased agricultural productivity. In fact, the district's response to technical change was almost on a par with that of Punjab and Haryana. In this area enough time has lapsed for the second-generation effects to manifest themselves and for their impact on the household incomes of the landless and near-landless to become evident.

Chapter 3 examines the impact of Green Revolution technology on the level, composition, and distribution of household income at the village level through a comparison of household incomes in the pre– and post–Green Revolution periods. The availability of a benchmark study for the mid-1960s for Walidpur village in Meerut District makes such a comparison possible. Our findings indicate that both

---

workers to engage in off-farm employment while remaining domiciled in their villages. Bhalla and Chadha (1981), in a study of Punjab, emphasized the growth linkage between higher rural incomes and the off-farm economy. They found that for every 1 percent increase in farm household income, clothing consumption increased by nearly 2 percent, footwear by 1.5 percent, and miscellaneous goods and services by around 1.3 percent. John Westley (1986), using aggregate state-level data for Punjab and Haryana, showed that poverty has been reduced in the two states as a result of agricultural growth. And John Harriss (1991), in a study on North Arcot District in Tamil Nadu, drew attention to the rising trend in the diversification of the rural economy.

total income and its distribution have improved with time, but that the distribution of crop income has worsened while that from non-crop sources has become better—implying that noncrop activities are important mechanisms in the income diffusion process.

Chapter 4 treats the three major mechanisms of income diffusion that operate in Meerut District: off-farm diversification, dairying, and the cultivation of labor-intensive, high-value crops. Chapters 5, 6, and 7 deal with each mechanism in the context of an individual case-study village. The criteria employed in the selection of these villages are also discussed in Chapter 4, as is the research methodology. The income profile of each village is constructed from a stratified random sample of households. Households are categorized on the basis of landowner-ship into landless, near-landless or marginal, small, medium, and large size groups. Total income per household is estimated by source—crop cultivation, agricultural labor, animal husbandry, off-farm employment, and remittances—and Lorenz curves illustrate the distribution of income from each source. A decomposition of the Gini coefficient identifies income sources that have a favorable effect on reducing inequality.

The investigation is carried one step further to selected households in order to understand how individual families have responded to the changing opportunities brought about by the second-generation ef-fects of the Green Revolution. The case-study households—described in the present tense although the particulars for them were collected in 1988–89—are all from the landless and near-landless categories. These households are not selected randomly; rather they are chosen from among the relatively prosperous families within the category. The individual case-study approach has some advantages. In the ab-sence of benchmark studies, the case studies provide an indication of changes brought about by technology and government intervention. Moreover, they help to identify some of the strategies adopted by successful households to overcome the constraints of limited land and capital resources. These strategies suggest possible areas where gov-ernment intervention might help to speed up the process of income diffusion.

Chapters 5, 6, and 7 deal with each mechanism discussed in Chap-ter 4. Using the methodology outlined above, data from each village demonstrate how the particular mechanism has been instrumental in diffusing income throughout the community.

Chapter 8 summarizes the findings of the village studies and draws policy implications from them. The strategies followed by successful landless and near-landless households are easily adaptable to a wider canvas and point to important changes that should be made in government's approach to rural development in India.

2

The Green Revolution in Uttar Pradesh: The Widening and Subsequent Narrowing of Regional Disparities

The regional impact of technical changes has varied over time in the state of Uttar Pradesh. The relative dynamism of western Uttar Pradesh during the initial phase of the post–Green Revolution period—the decade after the mid-1960s—can be traced to geographical and historical factors. The state's eastern region was at first barely affected. But, following improvements in its rural infrastructure, it too experienced rapid increases in productivity during the second phase, from the mid-1970s to the mid or late 1980s.

Profile of Uttar Pradesh

The state of Uttar Pradesh (U.P.) in northern India has had a significant influence on the social, economic, and political history of the country. Geography alone assures its position as the agricultural heartland. U.P. comprises almost half the Gangetic plain, one of the most fertile tracts in the world. It is the most populous state in India, with a population of more than 110 million in 1981. The population density of 377 persons per square kilometer is more than one and a half times the national average. Almost one out of every six Indians lives in U.P., although the state comprises only one-tenth of the country's land area (India, Office of the Registrar General, *Census of India*, General Population Tables).

Despite the wealth of its natural resources, U.P. remains one of the less developed states in the country. Per capita income in the mid-

1980s was about Rs 2,000, compared with the national average of Rs 2,600 (Uttar Pradesh, State Planning Institute, *Statistical Diary of Uttar Pradesh*). More than 80 percent of its population is rural, as against 75 percent for the nation. Three-fourths of the work force earns its living from the land, compared with two-thirds for India as a whole (Singh 1987). Landholdings are small and fragmented; in the mid-1980s the average holding was less than 1 hectare (Uttar Pradesh, Board of Revenue, *Agricultural Census of Uttar Pradesh*).

Through its sheer size and numbers U.P. dominates the agricultural scene of the nation. In the mid-1980s it accounted for about one-tenth of the net cultivated area in the country but one-fourth of the total irrigated area. Its contribution to total foodgrain production was over one-fifth, while its shares in wheat and sugarcane production were more than 35 and 40 percent, respectively (Uttar Pradesh, Department of Agriculture, *Uttar Pradesh Ke Krishi Ankre*).

Geographical Characteristics

About two-thirds of the state lies in the Gangetic plain, and 90 percent of its population lives there (Fig. 2.1). This highly fertile, intensively cultivated heartland incorporates three great rivers—the Ganges, the Yamuna, and the Ghagra—which rise in the Himalayas and flow roughly parallel to each other before they link near the state's eastern border.

Water Resources. Rainfall varies from an annual 130 centimeters in the north and northeast sections of the plain to 70 centimeters in the south and west (Fig. 2.2). In the extreme southwest, toward the Rajasthan desert, rainfall averages less than 70 centimeters. Nearly 80 percent of rainfall occurs during the June–September monsoon season. Irrigation is necessary for year-round cultivation to be practiced.

About 300 billion cubic meters of water flow through U.P. annually, of which about 125 billion cubic meters can be utilized for irrigation. In the mid-1980s less than one-third of this surface water was being exploited.

The groundwater resources of the Gangetic plain are equally abundant. The depth of the water table varies between 2 and 12 meters from ground level; the seasonal fluctuation ranges from 1 to 6 meters with an average of 3 meters. Both shallow and deep tubewells operate in this region. Of the existing groundwater resources, an estimated 38 percent were being exploited in the mid-1980s (Ramaseshan 1988).

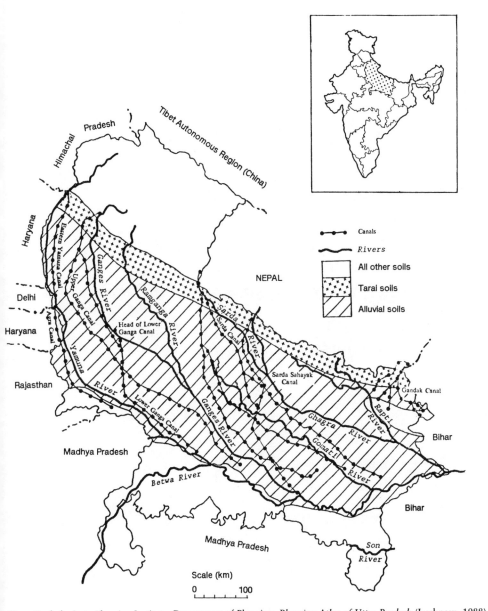

Source: Uttar Pradesh, State Planning Institute, Department of Planning, *Planning Atlas of Uttar Pradesh* (Lucknow, 1988).

Figure 2.1. Major rivers, canals, and soil groups in Uttar Pradesh

Regions and districts

HILL REGION

WESTERN REGION

CENTRAL REGION

EASTERN REGION

BUNDELKHAND REGION

Muzaffarnagar
Meerut
Ghaziabad
Bulandshahr
Bijnor
Badaun
Aligarh
Agra
Gorakhpur
Deoria
Faizabad
Varanasi

Rainfall

>200 cm
110–130 cm
90–110 cm
70–90 cm
<70 cm
• City

Scale (km)
0 50 100

Rainfall (cm)
80
70
60
50
40
30
20
10
0
J F M A M J J A S O N D
Months

DEHRA DUN 231.4
MEERUT 83.8
LUCKNOW 99.2
JHANSI 100.0
GORAKHPUR 127.4
VARANASI 111.3

Source: Uttar Pradesh, State Planning Institute, Department of Planning, *Planning Atlas of Uttar Pradesh* (Lucknow, 1988).

Flooding and waterlogging are major problems in the eastern portion of U.P. Deforestation in the upper catchment areas, resulting in soil erosion and silting of the riverbeds, is responsible for the floods that plague large tracts of the eastern region almost every year. Waterlogging is particularly acute in the Sharda Sahayak and Gandak irrigation command areas. In some tracts the water table has risen to less than 2 meters below ground level and the gain in irrigated area has been lost to waterlogging (Reserve Bank of India 1984).

Regional Divisions. The geographical components into which the state naturally divides are shown in Figure 2.2. The portion of the Himalayas falling within U.P. makes up the Hill region. The southwestern area extending into the central Indian plateau forms the Bundelkhand region, which derives its name from the Bundela Rajput kings who ruled its harsh terrain from the sixteenth to the eighteenth centuries. The third natural region, the Gangetic plain, can be further divided into western, central, and eastern zones, which differ in their history and economic status.

Our analysis of the impact of the Green Revolution is limited to the western and eastern regions, where the contrasts are most vivid. The western region comprises nineteen districts, while the eastern region has fifteen.[1]

Agrarian Structure

The western and eastern regions of Uttar Pradesh each account for about 30 percent of the total area of the state. They share equally in the population, each accounting for over one-third. Population density is also about the same in the two regions, around 480 persons per square kilometer (India, Office of the Registrar General, *Census of India*, General Population Tables). However, most parameters of agricultural performance indicate that the western region is more developed. Only 75 percent of the population in western U.P. is rural, compared with 90 percent in the east (Singh 1987). The relatively higher pressure on land in eastern U.P. has led to rapid fragmentation of landholdings. In 1980–81 there were about one and a half times as many holdings in the eastern region as in the west. The average hold-

1. Two new districts were created from existing ones in the western region and four in the eastern region in 1987–89. Separate official statistics for these districts are not available; hence, for the purposes of this study they are treated as part of their original districts.

ing was around 0.7 hectare in the east, compared with 1.2 hectares in the west (Vaish 1964).

Distribution of Landholdings. The distribution of landholdings by size is an important indicator of landownership, agricultural productivity, and demographic pressure on land resources. The evidence on landownership in western and eastern U.P. suggests that, contrary to the view that land concentration has increased over time, with big and medium farmers buying out small, uneconomic holdings, the equality of land distribution has improved. It is generally agreed that the "subdivision of land holdings and [the] sale and purchase of land seem to have been more important in bringing about . . . structural [re]distribution than the redistribution of land through tenancy reforms and legal ceilings on land holdings" (Rao 1976:61).

The pressure of population on the land can be gauged by the rapidly declining size of the average holding in the two regions, as indicated in Table 2.1.

In 1980–81 there were about 18 million landholdings in U.P., accounting for 18 million hectares of operated area. Of the total holdings, about 30 percent were in the western region and accounted for about 35 percent of the total area. The eastern region, on the other hand, had more than 42 percent of total holdings with less than 32 percent of the total operational area (Uttar Pradesh, Board of Revenue, *Agricultural Census of Uttar Pradesh*).

Changes in the distribution of number and area of holdings by size of landholding group are shown in Figure 2.3. In the western region, the percentage of marginal (<1 ha) holdings increased from more than 50 percent to 60 percent in the two decades following 1960–61. In the same period, the operational area under marginal holdings rose by approximately the same amount, implying that the average size of marginal holding remained almost constant. The category of small

Table 2.1. Average size of holdings in Uttar Pradesh, 1960–1961, 1970–1971, and 1980–1981 (ha)

Year	Western U.P.	Eastern U.P.
1960–61	1.65	1.40
1970–71	1.36	0.88
1980–81	1.22	0.74

Source: Uttar Pradesh, Board of Revenue, *Agricultural Census of Uttar Pradesh* (Lucknow, various years).

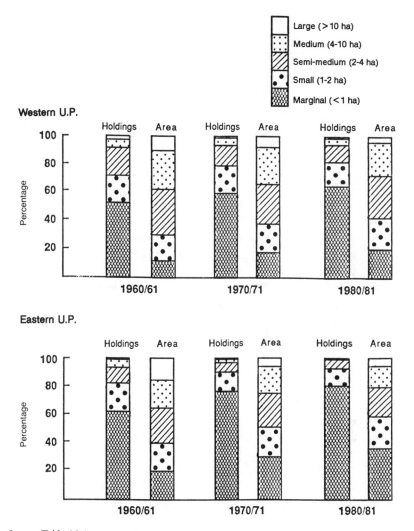

Source: Table A2.1.

Figure 2.3. Percentage distribution of landholdings and operational area in Uttar Pradesh, by size of landholding group, 1960–1961, 1970–1971, and 1980–1981

holdings maintained its share at around 20 percent and increased its area by 4 percent.

Eastern U.P. has a higher percentage of marginal holdings than the west. The proportion of marginal holdings increased from more than 60 percent to 80 percent between 1960–61 and 1980–81. The area under marginal holdings increased in the same ratio.

The increase in number of marginal holdings has been accompanied by a decrease in the number of medium (2–4 ha), semi-medium (4–10 ha), and large (>10 ha) holdings in both regions, and in eastern U.P. even the number of small holdings has declined. The class of small and medium farmers is smaller in eastern U.P. than in the west. Big farmers in both regions are fast disappearing.

Composition of Work Force. The rapidly declining size of holdings in both western and eastern U.P. suggests that the growing labor force is unable to find gainful employment outside agriculture. Figure 2.4 shows changes in the number of workers by region. In western U.P. the total number of workers was around nine million in 1970–71. Of these, 71 percent were engaged in agricultural activities, primarily crop cultivation and agricultural labor. In the decade of the 1970s the number of workers increased by two million. The share of agricultural workers declined to 69 percent—a shift of merely 2 percent out of agriculture.

In the eastern region the number of workers was higher, as also was the proportion of workers in agriculture—83 percent in 1970–71. During the next ten years the latter figure decreased to 79 percent, reflecting a shift of 4 percent out of agriculture. Even so more than nine million workers remained dependent on the land. Creation of nonagricultural employment opportunities during the 1970s was very sluggish.

Historical Perspective on East-West Disparities

The divergence in growth patterns between eastern and western U.P. did not occur suddenly as a result of the Green Revolution. In fact, as a review of the historical experience of the two regions illustrates, the imbalances that resulted from the introduction of new agricultural technology in the mid-1960s were in some ways a logical extension of a process that began more than a century ago.

As early as 1872, the east-west difference was noted by the Settlement Officer, who observed that "the laborer in Muzaffarnagar [western U.P.] dresses better than the average petty proprietor of the eastern districts, and wheat . . . forms a larger proportion of the food of the poorer classes" (Stone 1984:278).

In the eighteenth century the situation was quite the reverse. Describing the thriving economy of eastern Uttar Pradesh in the latter half of the eighteenth century, C. A. Bayly wrote (1983:104): "Be-

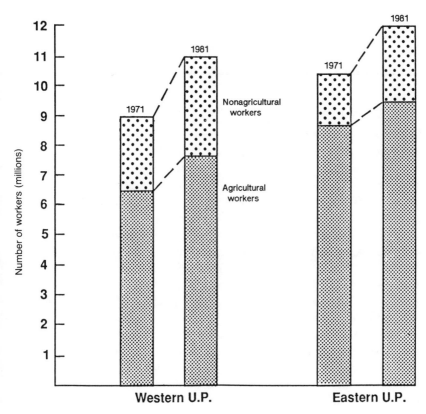

Source: Table A2.2.
Note: Nonagricultural includes animal husbandry, household industry, manufacturing, construction, trade, transport, and service; agricultural includes cultivators and agricultural laborers.

Figure 2.4. Composition of rural work force in Uttar Pradesh, by region, 1971 and 1981

nares [Varanasi] was one of the fastest growing cities during the years 1750–90. It became the subcontinent's inland commercial capital. . . . Yet it was also a city which benefitted from the sustained agricultural performance of the rich and stable tracts which surrounded it."

By the mid-nineteenth century the prosperity of the eastern region, based on export of such cash crops as sugar, indigo, opium, and rice and supplemented by a flourishing handloom textile industry, began to decline, giving way to the sugarcane and wheat belt of the western region and new manufacturing towns such as Kanpur. The decline in eastern U.P. also applied to east India. Commenting on the shifting scene of economic dynamism, Eric Stokes (1978:228) observed: "In this way, over the course of the nineteenth century Lakshmi, the fickle goddess of fortune, betook herself with uneven tread westward from the lush verdure of Bengal until she has come to fix her temporary abode on the Punjab plain between Ludhiana and Lyallpur."

Causes of Uneven Development

What determined the relative prosperity of the western region and the nature of the growth process that occurred there? It is sometimes argued that demographic pressures and landholding size have played pivotal roles in the east-west contrast, but the evidence does not support this hypothesis (Stokes 1978). Western U.P. was fairly well populated even before the spread of canal irrigation. In 1881 the population per cultivated square mile in Meerut division was 741, while in the districts of eastern U.P. it ranged from 914 in Gorakhpur division to 978 in Benares division (Stone 1984). While the population density in the western region was no doubt lower than in the more populous east, it is erroneous to believe that western U.P. was an unsettled tract until railways and canal irrigation unlocked its agricultural potential.

Stokes (1978) and Ian Stone (1984) argued forcefully that the major cause of regional imbalance was the prevailing land tenure system in the two regions and the development of canal irrigation in western U.P. but not in the east. In their view factors such as population pressure and the advent of railways played at best a supplemental role.

Disparities in the Land Tenure System. The impact of British land revenue settlements on eastern and western U.P. was fundamentally different. The eastern region, which came under British control a little earlier, experienced the Permanent Zamindari Settlement. High and inflexible revenue demands under this form of settlement led to the stratification of rural society into several layers of tenants, subtenants, and rentier landlords, almost completely alienating cultivators from the land. Land scarcity leading to rising rents made it possible for landowners and subtenants to rake in large surpluses in the form of rent. Subtenants were frequently traders and moneylenders who grew stronger under the *zamindari* system and forced the cultivator to sell his produce at a lower than market rate on account of indebtedness (Bhaduri 1985).

This situation contrasted with the *bhaiachara* system of the western districts, which was closely associated with owner-cultivator operations. This system led to greater security of tenure for cultivators and witnessed the rise of a class of rich occupancy tenants. The area under occupancy rights increased by 35,000 acres in Muzaffarnagar (western U.P.) between 1892 and 1920, prompting the Settlement Officer to comment: "This bespeaks a prosperous peasantry, for such valuable rights are normally allowed to accrue only with a substantial payment to the proprietor" (Stone 1984:308).

This peasant proprietorship was the crucial factor in the growth process of the western region. It created a class of independent tenants who had a much higher incentive to invest in land and improve productivity than did their counterparts in the east, which was reflected in the changes in crop patterns, improvement in yields, and capital accumulation (Stokes 1978). The pattern of use—even the very adoption—of so fundamental an input as canal irrigation was conditioned by tenure. Even within western U.P. the response of farmers varied between different land tenure systems. One canal engineer remarked upon the "rapid extension of irrigation in the villages owned by *Bhaiachara* communities [while] where the land belongs to large *zamindar* the increase, if any, is slight. Apparently cultivators in *zamindari* villages were afraid that landlords would use the advent of canal irrigation to enhance their rents and deprive them of *maurisi* rights" (Stone 1984:312).

While both regions were affected by demographic stress and the consequent fragmentation of holdings, the owner-cultivators adapted more readily to the need for increased production. In the east, insecure tenants crushed under the burden of exorbitant rents could not find the incentive to enhance productivity because the lion's share of their marketable surplus would be skimmed off by intermediaries. The rentier landlords, several times removed from the land, responded to the slowly deteriorating situation by trying to squeeze more from their impoverished tenants.

Critical Influence of Canal Irrigation. The introduction of canal irrigation into this uneven agrarian structure in the mid-nineteenth century further exacerbated the economic differences. The western region received a large amount of public investment in canal irrigation in the nineteenth century, and the eastern districts hardly any at all. In any case, when a proposal for a canal for the central and eastern region was first mooted in the 1870s it was opposed by the *talukdars* of Oudh, who were content to maintain the status quo (Whitcombe 1972). They relented much later, after the droughts and famine of the late nineteenth century began to result in agrarian unrest. As a result, the Sarda Canal was eventually completed in 1926 to cater to the central region (Clift 1977).

The eastern Yamuna, the Upper and Lower Ganga, and the Agra canals in western U.P. were built between 1830 and 1880. Canal irrigation triggered several changes in the agrarian economy, resulting in significant shifts from subsistence to cash crop production, primarily sugarcane and wheat in place of coarse cereals.

According to Stone (1984), whose history of canal irrigation in U.P. is a leading source on the subject, the opening of the canal systems was a major technical innovation. Canal waters spread faster and could irrigate larger tracts than the traditional wells, tanks, and ponds. The innovation allowed for an expansion in cash crop production by releasing both human and bullock labor from the more labor-intensive well and Persian wheel. The western peasants redirected the freed labor and bullock resources to more intensive use of the land through multiple cropping. The increased demand for labor was reflected in the wage rates of permanent farm servants and daily wage laborers, which were twice those prevailing in the eastern region around 1900. The Imperial Council of Agricultural Research Survey in 1939–40 found that adult males spent 145 days each year on crop production in Meerut as against 80 days in Gorakhpur (Stone 1984). The higher cost of labor provided an incentive for productivity-raising innovations. The efficiency of cane pressers, or *kolhu*, was improved, as was the quality of plows and draft cattle.

The boost to agricultural growth provided by the spread of canal irrigation was not limited to crop husbandry alone. A host of specialized nonagricultural activities also emerged in response to the demands of a more dynamic agricultural regime, leading to a qualitative change in the local economy. Villages outside the canal network, which could not participate directly in the new production systems, benefited indirectly by providing carting services for the enhanced volume of production. Other areas, less intensively cultivated, specialized in rearing draft cattle. Before 1908, fifteen flourishing factories producing various sizes of carts had been established in Meerut. "A large number of small towns" in Muzaffarnagar catered to the commercial needs of a prosperous peasantry (Stone 1984:301–302).

Growth of Infrastructure

The development of infrastructure spurred the growth process in western U.P. and was in turn stimulated by it. The evidence indicates that both regions were equally served by the railways, although the western region had better roads and market facilities.

Roads. Even at the turn of the century, Meerut District in western U.P. "was exceptionally well provided with railways and metalled roads"; thereafter they "improved considerably." In the 1930s there was not "a single village greater than fifteen miles from a railway station" (Stone 1984:320). Road improvements led to significant economies of scale. The situation of the roads in one of the major eastern districts was described by the Collector of Gorakhpur in 1882–83: "At least one-half of the grain which arrives in Gorakhpur arrives by river. The only metalled roads in the district are the military roads to Benares and Fyzabad. The *kutcha* roads are everywhere difficult and some of the most important of them are quite impossible for carts during the rains" (Stone 1984:315).

Market Facilities. There was a striking difference between the marketing systems of eastern and western U.P. The eastern region, characterized by a low marketable surplus, was served largely by periodic village markets (*painths*) and to a lesser extent by small market yards. Furthermore, poor communications prevented the development of the large-scale market yards, or *mandis*, that developed in western U.P. in response to the larger surplus. Marketing charges in the more efficient western *mandis*, where practices were standardized, were significantly lower than in the east. "The western districts therefore constituted an appropriate technical and institutional environment for maximizing peasant production . . . [and] guaranteed a widely spread distribution of income. . . . The stimulation to local expansion and diversification in general was related closely to the particular multiplier effects emanating from an agricultural sector characterized by small units" (Stone 1984:346).

Is History Repeating Itself?

We have reviewed the historical setting in some detail because certain aspects of agrarian history appear to be repeating themselves. The introduction of the HYV technology in the mid-1960s brought about an agricultural revolution in western U.P. in many ways similar to the earlier one. Both transformed the prevailing production system and were fueled by the irrigation process, the earlier one by canals, the later one by tubewell irrigation. While the first led to the substitution of wheat for coarse cereals and to the expansion of sugarcane as the major cash crop, the second shifted cropping patterns from pulses and oilseeds to HYVs of wheat and rice. As yields trebled and marketable surpluses increased, wheat attained the exalted status of a cash crop.

The most important consequence of the "first Green Revolution," besides setting the tone for commercialization of agriculture, was the initiation of the process of rural diversification. A hundred years later the same effects are becoming visible again, only magnified.

State Intervention to Correct Imbalances

On the eve of independence, Uttar Pradesh inherited a land tenure system that favored big landlords and *zamindars* at the expense of tenant cultivators, and also a canal irrigation system heavily biased in favor of the western region. In an attempt to redress the imbalances created by colonial policies, the new government undertook two major interventions: land reform measures were initiated to abolish the *zamindari* system, and the focus of irrigation was shifted from the relatively well developed districts of western U.P. to the eastern region. The salutary effects of these policies, especially irrigation, were not felt in eastern U.P. until almost the mid-1970s, by which time the Green Revolution had already gained a head start in the west.

Land Reform

The extreme inequality in landownership under the *zamindari* system can be gauged from the fact that big landlords, about 1.5 percent of the total proprietors, owned nearly 58 percent of the land in the state of Uttar Pradesh. Of these, about 800 of the biggest landlords, 0.001 percent of the total number, owned between one-third and one-fourth of all agricultural land. In 1950–51 almost all the agricultural land in the state, some 18 million hectares, was owned by *zamindars*, about 20 percent under their direct occupation and the remainder cultivated by tenants (Singh and Misra 1964).

The Uttar Pradesh Zamindari Abolition and Land Reforms Act was passed in 1951 to abolish all intermediary rights in land and to bring the actual cultivators into direct contact with the state. Multiple land tenures were replaced by a simplified uniform system whereby all cultivators were placed in one of two categories, *bhumidhars* (owners) and *sirdars* (hereditary tenants). In 1977, however, the latter category was abolished and all *sirdars* were given proprietary rights (Singh 1987, Singh and Misra 1964).

One of the major criticisms against the *zamindari* abolition was that the ex-*zamindars* were allowed to retain large areas of land that were (supposedly) under their direct cultivation, and with considerable manipulation the ex-landlords managed to inflate their possessions (Dantwala 1986a). Notwithstanding these shortcomings, the abolition of *zamindari* was a major step toward a more equitable agrarian structure.

Reorganizing Irrigation Priorities

In 1950–51 the acreage irrigated by canals in the western region was more than twelve times that in the East. In the subsequent period of planned development greater emphasis was laid on the exploitation of surface water in the eastern and central regions. With the development of two major irrigation projects, the Sharda Sahayak and the Gandak, canal irrigation improved significantly in eastern U.P. In fact, the ratio of canal irrigated area between the western and eastern regions declined from 12:1 in the early 1950s to about 5:1 in the early 1960s, and was further reduced to about 2.5:1 in the mid-1970s. By the mid-1980s the imbalance had diminished to a point where canal irrigation accounted for 1.2 and 0.9 million hectares in the western and eastern regions, respectively (Uttar Pradesh, Board of Revenue, *Crop and Season Reports of Uttar Pradesh*). By the time the eastern region caught up with the west in canal irrigation, however, the focus had already shifted to private tubewells, which make possible the more controlled and timely irrigation that is a prerequisite for the new technology.

Even as the state government tried to diffuse the availability of canal irrigation, a similar attempt was made with respect to state tubewells. The first state tubewells were installed in western U.P. in 1930. Their number continued to increase until by 1950 about 2,300 state tubewells were in operation, 96 percent of them in the western region. In the period following independence, far more state tubewells were installed in the eastern region so that the proportion of tubewells in the western region was reduced from 96 percent of the total in 1951 to 58 percent in the 1961, and to 54 percent in 1971. The share of the eastern region rose from around 4 percent in the early 1950s to 40 percent in 1971 (Clift 1977). The changing concentration of state tubewells in western and eastern U.P. is shown in Table 2.2.

The impact of government intervention in the form of rectifying the

Table 2.2. Number of state tubewells per
1,000 hectares of net sown area in Uttar
Pradesh, 1970–1971, 1980–1981, and 1985–
1986

Year	Western U.P.	Eastern U.P.
1970–71	0.98	0.77
1980–81	1.13	1.43
1985–86	1.50	1.70

Source: Uttar Pradesh, Board of Revenue,
Crop and Season Reports of Uttar Pradesh (Allahabad: Government Press, various years).

concentration of state tubewells lay not so much in the irrigation potential created—their contribution was a mere 8 percent of net irrigated area in the mid-1980s—as in the demonstrative effect of tubewell irrigation. This performed a crucial service in educating the
farmers in the methods and advantages of tubewell irrigation (Clift
1977). Thus, when the spread of modern varieties necessitated controlled irrigation, farmers in both regions readily installed private
tubewells.

Widening and Subsequent Narrowing of Regional Disparities

For the purpose of analysis the post–Green Revolution era is divisible into two time periods: the early phase, from the mid-1960s to
mid-1970s, and the later period, from the mid-1970s to the mid or
late 1980s. The early phase saw a widening of spatial disparities between the two regions, followed by a narrowing as the initially bypassed eastern region showed signs of accelerated agricultural growth.

Data at both the regional and district level can be used to illustrate
these changes. The regional data, however, must be interpreted with
caution because they tend to conceal differences within the region.
Therefore, wherever possible in what follows, the regional data are
accompanied by a map illustrating the situation in the individual districts that make up the region.

Agricultural Performance: Macro Evidence

The key indicators of agricultural performance for which data are
available are irrigation, cropping intensities, cropping patterns, agricultural productivity, and fertilizer use.

Irrigation. On the eve of the Green Revolution, irrigation coverage was almost equal in the two regions, with one major difference. Whereas in western U.P. more than half the acreage was irrigated by canals, in the eastern region more than 90 percent used water from traditional sources—wells, tanks, and ponds. With the advent of modern HYVs in the mid-1960s the very nature of irrigation began to change. Private tubewell irrigation gained in importance relative to other sources. The reason is obvious: HYVs of wheat and rice not only require greater quantities of water, they require it in controlled dosages. Canal irrigation does not have the flexibility that is possible with private tubewells; hence the dramatic rise in tubewell irrigation first in western U.P. and in the later period in the eastern region as well.

This transformation can be observed in Figure 2.5, which illustrates changes in net irrigated and tubewell irrigated area as percentages of the net cropped area. It is apparent that in the decade following the mid-1960s the western region underwent an impressive increase in irrigation—from roughly 40 to more than 60 percent of net cropped area—mainly as a result of expanded private tubewells. In the east the increase in irrigated area was marginal, although significant from the point of view of tubewell irrigation.

In the second period the irrigation status of eastern U.P. improved significantly—from 40 percent to more than 55 percent of the net cropped area. More significantly, the tubewell irrigated area rose from 15 percent to more than 35 percent of the net cropped area. In terms of absolute acreage, the irrigated area in western U.P. rose by 1.5 million hectares in the first phase, an increase three times that in the east. In the second phase, however, while an additional 0.77 million hectares were added in the west, the eastern region was close behind with 0.6 million hectares.

Thus, although eastern U.P. had a late start and still lags significantly behind the western region, both government intervention and private enterprise in exploiting its water resources led to a marked improvement in its irrigation status in the decade following the mid-1970s.

Cropping Intensities. Figure 2.6 indicates changes in cropping intensity—the indicator of the number of crops harvested on a piece of land during the year, calculated as gross cropped area divided by net cropped area—by region. It is clear that from the early 1960s to the mid-1960s the eastern region enjoyed a higher cropping intensity than the west. This is not surprising in view of the fact that the percentage

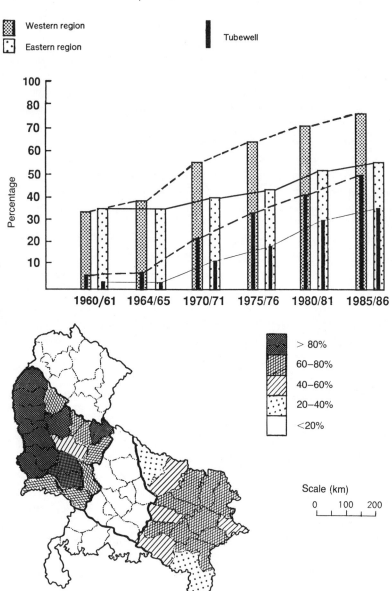

Sources: Table A2.3; Uttar Pradesh, State Planning Institute, Economics and Statistics Division, *Statistical Diary of Uttar Pradesh* (Lucknow, 1987).

Figure 2.5. Top, net irrigated area and tubewell irrigated area as percentage of net cropped area in Uttar Pradesh, by region, 1960–1961 through 1985–1986

Bottom, net irrigated area by districts of western and eastern regions as percentage of net cropped area, 1985–1986

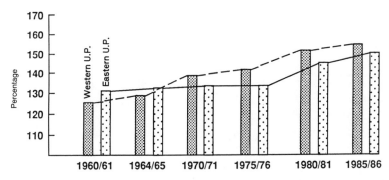

Source: Table A2.5.

Figure 2.6. Cropping intensity in Uttar Pradesh, by region, 1960–1961 through 1985–1986

of irrigated area at the time in both regions was almost equal; the eastern region with its higher rainfall during the monsoon was able to cultivate a larger *kharif* area. With the rapid increase of tubewell irrigation in western U.P. after the mid-1960s, the east lost its natural advantage.

In the first phase of the Green Revolution, western U.P. gained in terms of both *rabi* and *kharif* area and by the early 1970s had outpaced the east in double cropping. This trend continued well into the decade and after, with the cropping intensity registering impressive growth. Eastern U.P., having lost its initial advantage, remained almost static in the first phase, thereby widening the gap between the two regions. During the second phase, however, multiple cropping in the east picked up momentum as a result of the spurt in irrigation. Since then, the cropping intensity in the east has shown signs of catching up.

Cropping intensities in the two regions must be compared with caution. The official statistics treat sugarcane as a single *kharif* crop, but it remains on the field for an equivalent of two cropping seasons and thus should be counted as a double crop. The proportion of sugarcane in gross cropped area in the eastern and western regions has remained at around 10 and 4 percent, respectively. The cropping intensity is therefore slightly underestimated. Since the share of sugarcane remained almost unaltered over the period of comparison, however, the changes in trends can be considered fairly indicative of changes in multiple cropping.

Cropping Patterns. Shifts in cropping patterns reflect the response of the two regions to agricultural technology during the initial and later phases of the Green Revolution. Figure 2.7, showing the changing share of major crops in the gross cropped area, illustrates this.

WHEAT. Wheat has become the primary staple crop of western U.P. The first breakthrough in HYV technology came about in wheat, and the region was quick to adopt it.

The major shift in cropping patterns in western U.P. occurred in the early years of the introduction of the new technology. By the early 1970s not only was traditional wheat acreage replaced by HYVs, but competing crops such as *rabi* pulses and coarse cereals had lost ground to this important cereal crop. In the five years following the mid-1960s wheat acreage as a percentage of gross cropped area increased from around 20 percent to more than 30 percent while pulses and coarse cereals together registered an almost equal decline. Production of wheat, which stood at around 1.8 million tons in the mid-1960s, more than doubled between 1965 and 1970, registering a spectacular growth of about 15 percent per annum.

In the same period, the increase in wheat acreage in eastern U.P., although relatively lower, was significant. Production rose one and a half times in the first phase, from around 1 million tons to 1.5 million tons, a growth of about 7 percent per annum. While this was only half the growth achieved in the west, it was by no means unimpressive.

In the second phase, area under wheat in western U.P. increased marginally. Production increased from about 4 million tons in the early 1970s to around 7.5 million tons in the mid-1980s, implying a growth of some 4 percent per annum, considerably below its initial spurt.

The major impact of HYVs of wheat in eastern U.P. was experienced during the second phase when the region underwent a major transformation in its foodgrain-cropping pattern. Rice was formerly the predominant crop of the east, but with the advent of wheat technology the eastern region steadily increased its wheat acreage, especially during the second phase, so that by the mid-1980s wheat had attained a status almost equal to that of rice.

In the later period, wheat area increased impressively, from less than one-fifth to more than one-third of the gross cropped area. Production, which stood at around 1.5 million tons in the early 1970s, more than trebled, indicating a growth of about 8 percent per annum. It is significant that in the second phase the growth of wheat production in eastern U.P. was almost twice that in the west.

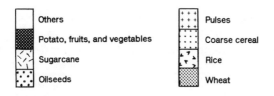

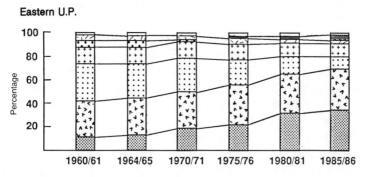

Western U.P.

Eastern U.P.

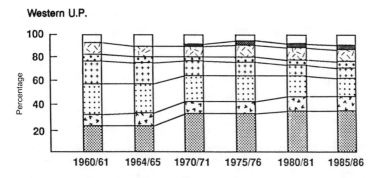

Source: Table A2.6.

Figure 2.7. Percentage area of major crops in gross cropped area in Uttar Pradesh, by region, 1960–1961 through 1985–1986

RICE. Although rice has been the traditional staple of the east, growth in this crop has not been as impressive as in wheat. In both regions its success has been limited in terms of area expansion and yield increases.

In the first phase, western U.P., with only one-tenth of gross cropped area under rice, did not experience any significant increase in acreage. Maize remained the major crop of the *kharif* season. The drier climate of western U.P. prevented major shifts into modern varieties of rice because of their much greater irrigation requirements. In areas of assured irrigation, however, the adoption of HYVs of rice led to increased yields and thereby enhanced overall production, which registered a growth of about 4 percent per annum.

Rice acreage remained almost static in eastern U.P. in the first phase. Productivity increases were also marginal. As a result, total output rose by a mere 1 percent per annum. The initial period therefore witnessed a widening disparity in the growth of output. In the second phase there was marked improvement in the irrigation status of the east. Adoption of modern varieties increased, and this was reflected in both acreage and production increases.

In western U.P. the second phase witnessed a growth rate of about 7 percent per annum in total rice production, which came about largely as a result of enhanced yields. The change in eastern U.P. was more dramatic. From a sluggish growth of about 1 percent in the first phase, production increased by an impressive 7 percent per annum in the second phase. While the major component of this growth was the contribution of enhanced productivity, there was an addition of about 0.3 million hectares in rice acreage as well. The improved performance of rice production in the second phase suggests that disparities are likely to be reduced in the future.

PULSES AND OILSEEDS. Pulses and Oilseeds were adversely affected by the Green Revolution in both the western and eastern regions. The early impact severely reduced acreage of *rabi* pulses and oilseeds, which compete with wheat. With growing emphasis on multiple cropping, long-duration pulses such as *arhar* (redgram) also diminished in importance (Sharma 1986).

In the latter half of the second phase, however, the declining trend in pulse acreage began to be reversed. This reflects the development of very short-duration varieties of *mung* (greengram) and *urd* (blackgram), which do not compete with the major cereal crops and are cultivated in intercropping sequences or as bonus crops between regular cropping seasons. With improving irrigation facilities this trend is likely to increase.

VEGETABLES, FRUITS, AND POTATOES. Although not very significant in the first phase, vegetables, fruits, and potatoes are emerging as alternate cash crops. Rising incomes and an increasing demand for fresh produce are likely to give these nontraditional cash crops a bigger role in agriculture in the future.

Agricultural Productivity. Yields are the usual indicators of agricultural productivity, but the yields of wheat and rice do not illustrate the lag effect between the two regions as vividly as some of the previous indicators (Fig. 2.8). One possible explanation is the sensitivity of yields to changes in climatic conditions. In the early phase, when the irrigation coverage was less, years of poor rainfall—such as 1974–75–significantly lowered average wheat yields. Indicators based on acreage were not affected to the same degree; hence the lag effect shows up more boldly in irrigated acreage, cropping intensity, and cropping patterns.

A more sensitive indicator of changing agricultural productivity is the monetary value of total output per hectare. The regional and district-level data shown in Figures 2.9, 2.10, and 2.11 were compiled from Bhalla and Tyagi (1989a). Figure 2.9 indicates the changes in value of agricultural output per hectare by region. In both western and eastern U.P. the growth in productivity was higher in the second period than in the first, but again the lag effect is not very evident. It

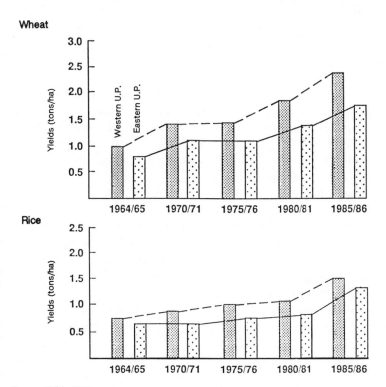

Source: Table A2.7.

Figure 2.8. Yields of wheat and rice in Uttar Pradesh, by region, 1964–1965 through 1985–1986

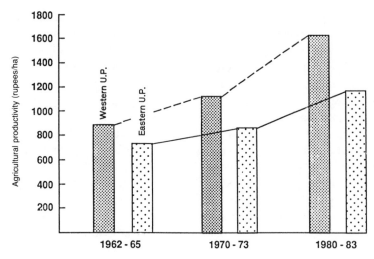

Source: Table A2.8.

Figure 2.9. Value of agricultural output per hectare for total crop sector in Uttar Pradesh, by region, 1962–1965 through 1980–1983 (at constant prices, average of 1967–1970)

is obvious, however, if district-level data are considered (Fig. 2.10). Bhalla and Tyagi classified the districts into five categories on the basis of their productivity level. We have simplified these into three categories. During 1962–65, the pre–Green Revolution period, in western U.P. 16 percent of districts were in the high–productivity category, 63 percent were in the medium category, and 20 percent were in the low-productivity category. In eastern U.P., by contrast, there was not a single district in the high-productivity class; 60 percent were in the medium and 40 percent were in the low category.

The second time period, 1970–73, coincides with the initial phase of the post–Green Revolution era. It is apparent from Figure 2.10 that while progress in western U.P. was impressive, the districts of the east remained almost static at the level of the mid-1960s. The number of high-productivity districts in western U.P. doubled to 32 percent, while all the low-productivity districts "graduated" into the medium class. In eastern U.P. the situation improved only marginally, with only two of the six low-productivity districts moving into the medium category.

By 1980–83, the second phase of the post–Green Revolution period, there was considerable progress in eastern U.P. Three districts moved into the high-productivity class, the medium-level districts ex-

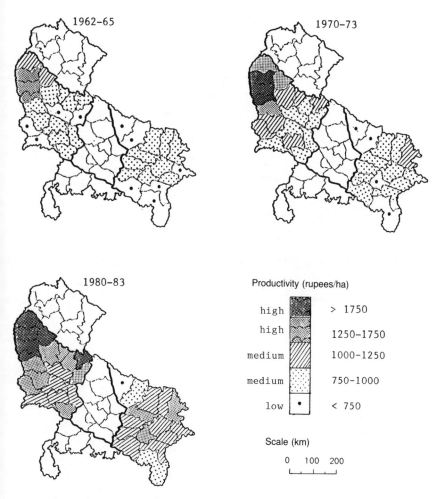

Source: Compiled from G. S. Bhalla, and D. S. Tyagi, *Patterns in Indian Agricultural Development: A District Level Study* (New Delhi: Institute for Studies in Industrial Development, Indraprastha Estate, 1989).

Figure 2.10. Value of agricultural output per hectare of major crops in Uttar Pradesh, by districts of western and eastern regions, 1962–1965, 1970–1973, and 1980–1983 (at constant prices, average of 1967–1970)

hibited considerable improvement, and only one district lagged behind in the low category. This acceleration of agricultural growth is brought out vividly in Figure 2.11, which displays the annual compound growth rates of total agricultural output in the districts of western and eastern U.P. In the initial period nearly 60 percent of the

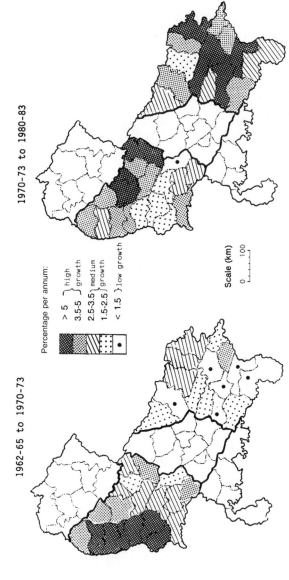

1962-65 to 1970-73

1970-73 to 1980-83

Percentage per annum:

> 5 } high growth
3.5-5 } growth
2.5-3.5 } medium growth
1.5-2.5 } growth
< 1.5 } low growth

Scale (km)

0   100

Source: Compiled from G. S. Bhalla and D. S. Tyagi, *Patterns in Indian Agricultural Development: A District Level Study* (New Delhi: Institute for Studies in Industrial Development, Indraprastha Estate, 1989).

Figure 2.11. Growth rates of total value of agricultural output of major crops in Uttar Pradesh, by districts of western and eastern regions, 1962–1965 to 1970–1973 and 1970–1973 to 1980–1983

48

districts of western U.P. were in the high-growth category, and half registered a growth of more than 5 percent per annum. In the eastern region, by contrast, 40 percent of districts were in the low-growth category and registered a growth rate of less than 1.5 percent per annum.

In the second period, however, the agricultural performance registered a remarkable change. In the western region four of the very high growth districts of the previous period, unable to sustain the pace, slumped into the medium category, and of the eight districts in the medium category five moved up into the high category. But in eastern U.P. the impact of growth was more spectacular. Nearly 75 percent of the districts moved into the high category, and more than half of these recorded growth rates higher than 5 percent per annum. The low-productivity class completely disappeared. The second phase clearly witnessed a catching up by the eastern region.

Trends in Fertilizer Use. The increasing use of chemical fertilizer is an indication of the spread of technology. In the pre–Green Revolution period chemical fertilizer use was negligible in both the western and eastern regions of the state (Fig. 2.12). Consumption of chemical fertilizers per hectare of gross cropped area increased severalfold in the decades following the introduction of the modern varieties.

In the western region it increased from about 6 kilograms per hectare in the mid-1960s to almost 100 kilograms in the mid-1980s. During the same period in eastern U.P. the increase was from 4 kilo-

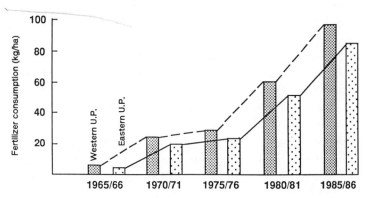

Source: Table A.29.

Figure 2.12. Fertilizer consumption per hectare of gross cropped area in Uttar Pradesh, 1965–1966 through 1985–1986

grams to about 85 kilograms per hectare. The rise in fertilizer consumption has been almost equal in the two regions; however, in the second phase its usage has increased most impressively.

The Growing Importance of Rural Growth Centers

It is clear that the impact of Green Revolution technology, first felt in the western region, has now extended its spatial coverage to the eastern region. The later period of the post–Green Revolution era is also characterized by the emergence of some second-generation effects, prominent among them the diversification of the rural economy into off-farm activities. These off-farm activities are contributing to the emergence of small rural towns and growth centers, which are increasingly the locus of activities in the trade, service, transportation, and small manufacturing sectors.

We would expect to find evidence of such changes in a greater increase in urban population in the post–Green Revolution period, a rise in the number and growth of small and medium-sized towns, and the wider dispersal of small towns or growth centers in the western region, where the agricultural technology has had the longest gestation and where its second-generation effects are most likely to be visible.

Although major changes in the growth of small and medium-sized towns have taken place in the decade of the 1980s, these will have to await the report of the 1991 census. On the basis of available statistics, however, trends in the growth of small towns and urban population appear to be consistent with our expectations.

Figure 2.13 indicates that while both urban and total populations increased at almost the same rate in the 1960s, in the following decade the urban population grew much faster. The acceleration in urban growth was experienced almost equally in the western and eastern regions. The degree of urbanization was much higher in the western region, where the absolute urban population continued to remain almost twice that in the east. The proportion of urban to total population in western U.P. rose from 17 to 24 percent between 1961 and 1981, and it increased from 7 to 11 percent in eastern U.P.

A more striking change in the trend of urbanization is the growth and proliferation of small and medium-sized towns in the 1970s. A town, according to the census definition, must have three-fourths of its working population dependent on nonagricultural pursuits (India,

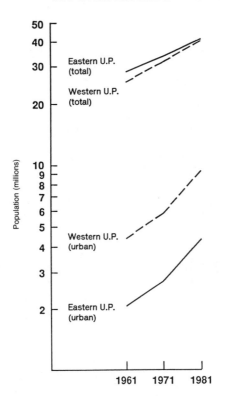

Source: India, Office of the Registrar General, *Census of India,*
General Population Tables, Series 1 (Delhi, various years).

Figure 2.13. Growth of total and urban population in Uttar Pradesh, by region,
1961–1981 (logarithmic vertical scale)

Office of the Registrar General, *Census of India,* General Population
Tables).

The growth of urban population in Uttar Pradesh by class of town
is shown in Figure 2.14. While population in Class I towns increased
marginally in the decade of the 1970s, in some of the smaller (Class
IV, V, and VI) towns it rose much faster. This implies an increase in
nonagricultural activities in the rural hinterland where most small
towns are located. It also suggests the beginning of a healthy trend in
the decentralization of urban population from a few large cities to
more widely dispersed smaller urban centers.

Figure 2.15 shows the location of various categories of towns in the

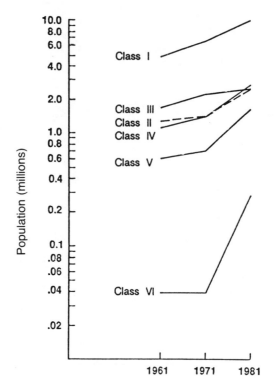

Class I:	100,000 and above	Class IV:	10,000-19,999
Class II:	50,000-99,999	Class V:	5,000-9,999
Class III:	20,000-49,999	Class VI:	< 5,000

Number and population of towns by class

	1961 number	Percentage of population	1971 number	Percentage of population	1981 number	Percentage of population
Class I	17	54.4	22	57.1	30	51.5
Class II	16	11.8	20	10.3	37	12.7
Class III	52	16.7	67	16.7	85	12.3
Class IV	75	11.0	91	10.4	194	13.4
Class V	74	5.9	80	4.7	231	8.6
Class VI	10	0.2	13	0.2	82	1.5
All classes	244	100	293	100	659	100

Source: India, Office of the Registrar General, *Census of India*, General Population Tables, Series 1 (Delhi, various years).

Figure 2.14. Urban growth in Uttar Pradesh, 1961–1981
Growth of population by class of town (logarithmic vertical scale)

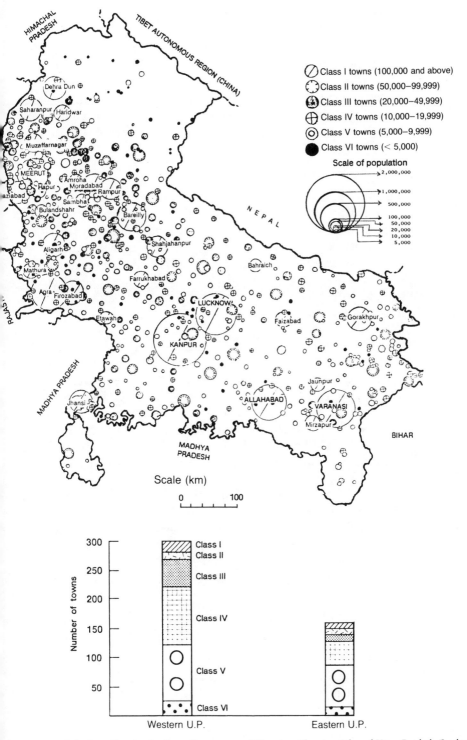

Class I towns (100,000 and above)
Class II towns (50,000–99,999)
Class III towns (20,000–49,999)
Class IV towns (10,000–19,999)
Class V towns (5,000–9,999)
Class VI towns (< 5,000)

Scale of population

2,000,000
1,000,000
500,000
100,000
50,000
20,000
10,000
5,000

Scale (km)

0   100

rce: Uttar Pradesh, State Planning Institute, Department of Planning, *Planning Atlas of Uttar Pradesh* (Lucknow, *8*).

ure 2.15. Geographical dispersion of towns in Uttar Pradesh, by class, 1981

state. As expected, small and medium-sized towns are widely dispersed in western U.P., suggesting that there might be a relationship between off-farm diversification brought about by the second-generation effects of the Green Revolution and the proliferation in the number and population of growth centers. Other studies in Punjab and Haryana seem to confirm this trend (Bala 1986, Chadha 1983).

Development of Infrastructure

The supposition that the extent and timing of economic changes has been chiefly the result of technical change in agriculture is further confirmed by an examination of the state's rural infrastructure. One of the oft-cited reasons for regional disparities in growth is disparities in their infrastructural endowments. A review of three major infrastructure facilities in U.P.—rural electrification, institutional credit, and roads—indicates few differences in extent and growth between the east and the west.

Rural Electrification. Rural electrification is a vital input for the modern production process. The cost of running tubewells on electricity is considerably lower than diesel power. The use of electricity in agriculture, primarily for tubewell irrigation, has been increasing rapidly. In the early 1960s the percentage of electricity consumed in agriculture in U.P. was about one-fifth of total consumption (National Council of Applied Economic Research 1965); by the mid-1980s this had increased to one-third (Uttar Pradesh, State Planning Institute, *Statistical Diary of Uttar Pradesh*). The available information suggests that coverage is improving in both regions, and the east is only slightly behind the west.

In the early 1980s the proportion of villages electrified to the total number of villages in the western and eastern regions was about 55 and 45 percent, respectively (Singh 1987). By the mid-1980s these figures had increased to 74 and 63 percent, respectively (Uttar Pradesh, State Planning Institute, *Statistical Diary of Uttar Pradesh*). There was, however, a considerable difference in consumption of electricity between the two regions.

Institutional Credit. Disparities in commercial banking facilities between the two regions were significant in the early 1950s, but the gap has been narrowing ever since. In 1951, for every one million persons there were more than ten bank branches in western U.P. and less than four in the east. In the early 1980s the number of banks in the two regions had risen to around twenty-five and twenty, respectively, per million people.

It is, however, the cooperative credit structure, and not the commercial banks, that constitutes the major source for agricultural credit in U.P. The short-term loans disbursed by cooperatives during 1982–83 were Rs 750 million in western U.P. (Rs 122 per net cropped hectare) as against Rs 530 million (Rs 94 per hectare) in eastern U.P. (Reserve Bank of India 1984). This is an improvement over the early 1970s when the credit disbursed by Primary Agricultural Credit Societies in western U.P. per unit of cropped area was about one and a half times higher than in eastern U.P. (Singh 1981). Official statistics do not provide an indication of differences in credit disbursement between various rural groups.

Roads. In 1961 there were about 23,000 kilometers of paved roads in U.P. The eastern region, with 30 percent of the land area and 38 percent of the population, was served by 30 percent of the total road mileage. The western region, with 31 percent of the area and 35 percent of the population, had 33 percent of the road mileage (National Council of Applied Economic Research 1965). By the early 1960s any imbalances in road network that may have existed prior to independence were rectified. In the mid-1970s the total length of roads in U.P. had increased to 37,000 kilometers, with each of the two regions still maintaining about one-third (Shankar 1978). In the mid-1980s the total length of roads increased to 52,000 kilometers (Uttar Pradesh, State Planning Institute, *Statistical Diary of Uttar Pradesh*). In the two regions the road length per 1,000 square kilometers increased from about 30 kilometers to 100 kilometers in the period from the early 1960s to the mid 1980s. Thus, insofar as the critical variable of road infrastructure is concerned, eastern and western U.P. have remained on a par.

Income Differentials: Evidence from Micro Studies

The impact of the Green Revolution on income disparities between the two regions is difficult to evaluate on the basis of macro-level data. There are two statistics published officially with regard to income: per capita state income and per capita net output from commodity-producing sectors at the district level. The former is an indicator of the total economy, and the latter includes five commodity-producing sectors only and excludes services. It is not possible to compile regional-level estimates from these data. But the information in Table 2.3 (in constant 1970–71 prices) for Meerut and Varanasi districts may be suggestive of differences between western and eastern U.P.

Per capita net output for Meerut was nearly twice as high as for Varanasi in the early 1970s, and while both districts registered an increase in real terms through the mid-1980s, there is no evidence of the difference narrowing. While this statistic provides a broad indication of the order of income differentials between the two regions, it sheds no light on changes in income distribution between various groups within each region. The only evidence available to demonstrate this aspect of the impact of Green Revolution technology is from farm management and other micro-level studies.

Initial Indicators of Growing Inequality

More studies of income distribution—perhaps as many as a dozen—are available for western U.P. than for the east. The early micro-level studies tended to corroborate what was then the prevailing wisdom: in both regions the Green Revolution appeared to have had an adverse effect on income distribution in the initial period.

One study compared farm business income in the mid-1950s and late 1960s in Muzaffarnagar District, based on extensive farm management survey data. Lorenz curves and Gini coefficients in the two periods indicated that the distribution of farm income had worsened over time (Saini 1976a). In another study, S. L. Shah and R. C. Agrawal (1970) analyzed the income levels of "progressive" and "less progressive" farmers in Badaun District in 1967–68. Progressive

Table 2.3. Per capita state income and per capita net output for Meerut and Varanasi districts, Uttar Pradesh, 1970–1971, 1980–1981, and 1984–1985 (rupees)

Year	Per capita state income U.P.	Per capita net output	
		Meerut	Varanasi
1970–71	486	474	255
1980–81	519	501	294
1984–85	585	553	302

Source: Uttar Pradesh, State Planning Institute, *Statistical Diary of Uttar Pradesh* (Lucknow, various years).

farmers were defined with the criteria of irrigated land, use of HYVs, chemical fertilizers, and ownership of means of irrigation. Within each category farmers were further classified into small, medium, and large on the basis of the amount of land they owned. Income from nonagricultural as well as agricultural sources was estimated for a sample of about 100 farmers each in the "progressive" and "less progressive" categories. There was a significant difference in the income levels of the "progressive" and "less progressive" farmers in the different size groups. The authors concluded that while the new technology increased the income levels of the medium and large farmers, the small farmers did not gain as much from it. As for income distribution, it was more unequal between the various size groups within the "progressive" category than the "less progressive," suggesting an unfavorable effect of modern technology.

His study of eight villages in Muzaffarnagar District led Pranab Bardhan (1970) to conclude that the economic condition of agricultural workers had worsened as a consequence of technical change. In seven of the villages real wage rates for casual male laborers declined between 1954–55 and 1967–68.

Subsequent Signs of Improvement

In one of the few studies in Uttar Pradesh in the later period that drew attention to the emerging second-generation effects of the Green Revolution, Etienne (1988) carried out an extensive survey of several regions where the Green Revolution technology had had varying impacts. Changes in the rural economy of a village in Bulandshahr District (western U.P.) were compared with those of a village in Varanasi District (eastern U.P.).

In Bulandshahr District, Etienne found that the process of agricultural growth had led to diversification of the rural economy, as evident in the increase in off-farm activities in the transport, business, and service sectors. The block headquarters town experienced a proliferation of workshops, garages, tea stalls, and the like. Trade and commerce were booming as demand for the services of tailors, cloth merchants, mechanics, electricians, photographers, and barbers increased. Domestic small industries and handicrafts, as well as the construction trade, provided employment for an increasing number of rural laborers. Most landowners in the study village with holdings of 1 hectare or more had replaced their mud houses with brick construction.

Etienne said the Varanasi District had entered a "transitional phase and [was] moving towards faster growth." The experience in Varanasi District was being replicated in other parts of eastern U.P. He concluded: "The long period of a semi-stagnant economy is over as a result of more dynamic trends: in particular, tubewells, rising consumption of chemical fertilizers and new seeds" (Etienne 1988:111).

Ground breaking though it is, Etienne's study is more qualitative than quantitative in nature and says little about the actual impact of noncrop activities on income diffusion and the well-being of the rural poor. This is the subject of the remainder of the present volume. The secondary effects of technical change are highlighted by demonstrating how they have led to the diversification of employment opportunities for landless and near-landless households. To illustrate this emerging trend, one must have a record of past circumstances against which to compare it. Such a benchmark is available for Walidpur village in Meerut District.

Appendix

Table A2.1. Percentage distribution of landholdings and operational area, by landholding group, in Uttar Pradesh, 1960–1961, 1970–1971, and 1980–1981

Size group	Western U.P.						Eastern U.P.					
	1960–61		1970–71		1980–81		1960–61		1970–71		1980–81	
	Holdings	Area	Holdings	Area	Holdings	Area	Holdings	Area	Holdings	Area	Holdings	Area
Marginal (<1 ha)	52.6	11.1	59.4	17.1	62.6	20.2	62.0	18.7	75.4	27.7	79.3	34.4
Small (1–2 ha)	19.1	17.6	19.7	20.5	19.3	22.1	19.0	20.6	13.9	21.8	12.6	23.4
Semi-medium (2–4 ha)	18.7	32.8	13.8	28.2	12.3	28.1	12.3	24.7	7.4	22.7	5.8	21.3
Medium (4–10 ha)	8.2	27.6	6.4	26.2	5.3	24.6	5.1	20.2	2.9	18.7	2.0	15.6
Large (>10 ha)	1.5	10.9	0.7	7.6	0.4	4.9	1.7	15.7	0.5	9.1	0.3	5.4
TOTAL	100	100	100	100	100	100	100	100	100	100	100	100

Sources: Uttar Pradesh, Board of Revenue, Agricultural Census, 1970/1971, 1980/1981 (Lucknow). The 1960–61 data are from R. N. Tewari, Agricultural Development and Population Growth: An Analysis of Regional Trends in U.P. (Delhi: S. Chand and Sons, 1970), p. 198, and are based on National Sample Survey estimates; they may not be strictly comparable with the later census figures.

Table A2.2. Composition of Uttar Pradesh work force, by region, 1971 and 1981 (millions)

Workers	Western U.P.		Eastern U.P.	
	1971	1981	1971	1981
Agricultural	6.40	7.63	8.63	9.39
	71.2%	69.1%	82.7%	78.9%
Nonagricultural	2.59	3.41	1.80	2.51
	28.8%	30.9%	17.3%	21.1%
TOTAL	8.99	11.04	10.43	11.90
	100%	100%	100%	100%

Source: India, Office of the Registrar General, *Census of India*, General Population Tables, Series 1 (Delhi, various years).
Note: Agricultural includes cultivators and agricultural laborers; nonagricultural includes workers in animal husbandry, household industry, manufacturing, construction, trade, transport, and services.

Table A2.3. Net irrigated area as percentage of net cropped area in Uttar Pradesh, by region, 1960–1961 through 1985–1986

Region	1960–61	1964–65	1970–71	1975–76	1980–81	1985–86
Western U.P.	34.3	38.7	56.2	65.3	72.4	77.4
Eastern U.P.	35.5	36.3	40.8	44.0	52.7	56.2

Sources: Compiled from Uttar Pradesh, Board of Revenue, *Crop and Season Reports of Uttar Pradesh* (Allahabad: Government Press, various years); Uttar Pradesh, State Planning Institute, Economics and Statistics Division, *Statistical Diary of Uttar Pradesh* (Lucknow, various years).

Table A2.4. Tubewell irrigated area in Uttar Pradesh as percentage of net irrigated area (NIA) and net cropped area (NCA), by region, 1960–1961 through 1985–1986

Region	1960–61		1964–65		1970–71		1975–76		1980–81		1985–86	
	NIA	NCA	NIA	NCA	NIA	NCA	NIA	NCA	NIA	NCA	NIA	NCA
Western U.P.	17.6	6.1	20.1	7.8	40.9	23.0	52.0	34.0	59.0	42.7	64.6	50.0
Eastern U.P.	8.3	2.9	10.8	3.9	31.9	13.1	42.0	18.4	59.1	31.2	61.0	34.3

Sources: Compiled from Uttar Pradesh, Board of Revenue, *Crop and Season Reports of Uttar Pradesh* (Allahabad: Government Press, various years); Uttar Pradesh, State Planning Institute, Economics and Statistics Division, *Statistical Diary of Uttar Pradesh* (Lucknow, various years).

Table A2.5. Cropping intensity in Uttar Pradesh, by region, 1960–1961 through 1985–1986 (percentage)

Region	1960–61	1964–65	1970–71	1975–76	1980–81	1985–86
Western U.P.	127.1	129.3	139.4	141.5	150.7	154.2
Eastern U.P.	131.7	132.8	134.0	134.0	145.2	150.0

Sources: Uttar Pradesh, Board of Revenue, *Crop and Season Reports of Uttar Pradesh* (Allahabad: Government Press, various years); Uttar Pradesh, State Planning Institute, Economics and Statistics Division, *Statistical Diary of Uttar Pradesh* (Lucknow, various years).

Table A2.6. Percentage area of major crops in gross cropped area in Uttar Pradesh, by region, 1960–1961 through 1985–1986

	Western U.P.						Eastern U.P.					
Crop	1960–61	1964–65	1970–71	1975–76	1980–81	1985–86	1960–61	1964–65	1970–71	1975–76	1980–81	1985–86
Wheat	22.3	21.6	31.6	31.0	34.0	33.0	11.7	12.9	18.2	22.2	31.7	33.1
Rice	10.1	10.5	10.5	11.1	12.3	12.5	31.1	32.1	31.4	32.4	34.6	35.0
Coarse cereals	26.0	24.9	23.1	23.4	19.4	17.3	28.7	27.5	29.6	21.0	13.5	10.0
Pulses	18.5	18.0	11.5	9.3	7.4	7.9	16.5	15.0	11.2	13.5	11.2	11.1
Total foodgrains	76.9	75.0	76.7	74.8	73.1	70.7	88.0	87.5	90.4	89.1	91.1	89.7
Oilseeds	5.0	5.0	3.4	4.5	3.9	4.7	5.0	4.5	1.5	1.4	1.5	1.5
Sugarcane	10.4	10.3	9.3	10.9	9.8	10.3	4.4	4.2	4.2	4.6	3.3	3.3
Potato	0.6	0.6	0.8	1.0	1.4	1.5	0.5	0.5	0.8	0.9	1.1	1.1
Vegetables and fruits	—	1.0	1.0	1.5	2.0	2.7	—	0.6	0.6	0.8	1.0	3.1
Others	7.1	8.1	8.8	7.3	9.8	10.1	2.1	2.7	2.5	3.2	3.0	1.1
Gross cropped area	100	100	100	100	100	100	100	100	100	100	100	100

Sources: Uttar Pradesh, Board of Revenue, Crop and Season Reports of Uttar Pradesh (Allahabad: Government Press, various years); Uttar Pradesh, Department of Agriculture, Directorate of Agricultural Statistics and Crop Insurance, Uttar Pradesh Ke Krishi Ankre [Agricultural statistics of Uttar Pradesh] (Lucknow, various years).

Table A2.7. Yields of wheat and rice in Uttar Pradesh, by region, 1964–1965 through 1985–1986 (tons/ha)

	Western U.P.		Eastern U.P.	
Year	Wheat	Rice	Wheat	Rice
1964–65	1.0	0.7	0.8	0.7
1970–71	1.4	0.9	1.1	0.7
1975–76	1.3	1.0	1.1	0.8
1980–81	1.8	1.2	1.3	0.8
1985–86	2.5	1.8	1.8	1.4

Sources: Compiled from Uttar Pradesh, Board of Revenue, Crop and Season Reports of Uttar Pradesh (Allahabad: Government Press, various years); Uttar Pradesh, Department of Agriculture, Directorate of Agricultural Statistics and Crop Insurance, Uttar Pradesh Ke Krishi Ankre [Agricultural statistics of Uttar Pradesh] (Lucknow, various years).

Table A2.8. Value of agricultural output per hectare for total crop sector in Uttar Pradesh, by region, 1962–1965 to 1980–1983 (rupees/ha; 1967–70 constant prices)

Region	1962–65	1970–73	1980–83
Western U.P.	899	1,125	1,614
Eastern U.P.	736	859	1,151

Source: Compiled from G. S. Bhalla and D. S. Tyagi, Patterns in Indian Agricultural Development: A District Level Study (New Delhi: Institute for Studies in Industrial Development, Indraprastha Estate, 1989), pp. 84–94.

Table A2.9. Fertilizer consumption per hectare of gross cropped area in Uttar Pradesh, by region, 1965–1966 through 1985–1986 (kg/ha)

Region	1965–66	1970–71	1975–76	1980–81	1985–86
Western U.P.	6.0	22.7	27.0	58.5	94.6
Eastern U.P.	4.2	19.6	22.6	49.6	82.9

Sources: Ajit Kumar Singh, Agricultural Development and Rural Poverty (New Delhi: Ashish Publishing House, 1987); Uttar Pradesh, Department of Agriculture, Directorate of Agricultural Statistics and Crop Insurance, Uttar Pradesh Ke Krishi Ankre [Agricultural statistics of Uttar Pradesh] (Lucknow, various years).

3

Income Diffusion over Time:
Walidpur Village

Walidpur, in Meerut District of western Uttar Pradesh, is our site for observing the impact of the Green Revolution on income diffusion at the village level. An average-sized, unobtrusive village, Walidpur demonstrates significant changes in the level and composition of household income as a result of Green Revolution technology. As the rural economy has diversified, employment opportunities in noncrop activities have increased. A large number of the village poor have participated in these activities, which are characterized by their low capital but high labor requirements. This participation is reflected in the rising share of noncrop earnings in the total household income of the landless and near-landless households. By comparing the nature and distribution of household income in the pre– and post–Green Revolution periods, we can show that there has been an improvement in both absolute and relative incomes of the rural poor in Walidpur village.

For such a time comparison we needed a benchmark study of the pre–Green Revolution period. This was the prime consideration for selecting Walidpur as the study village. The Agro-Economic Research Center (AERC) of Delhi University conducted a socioeconomic survey of Walidpur in 1958–59 (Vaish 1964), followed by another in 1963–64, and yet another in 1983–84 (Tyagi 1988). The surveys provide very detailed and excellent accounts of the socioeconomic conditions of the village. For purposes of comparison, we used income data from the 1963–64 study.

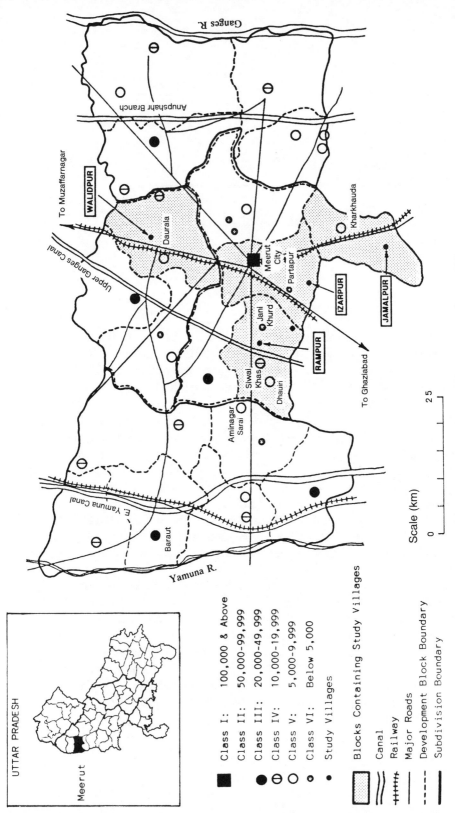

UTTAR PRADESH

Meerut

Class I: 100,000 & Above
Class II: 50,000-99,999
Class III: 20,000-49,999
Class IV: 10,000-19,999
Class V: 5,000-9,999
Class VI: Below 5,000
              Study Villages

Blocks Containing Study Villages

Canal
Railway
Major Roads
Development Block Boundary
Subdivision Boundary

Scale (km)

0   2   5

Source: Uttar Pradesh, State Planning Institute, Economics and Statistics Division, *Sankhyiki Patrika, Meerut* (Lucknow, 1987).

Ganges R.

Anupshahr Branch

To Muzaffarnagar

WALIDPUR

Daurala

Upper Ganges Canal

Meerut City

Kharkhauda

Partapur

IZARPUR

JAMALPUR

RAMPUR

Jani Khurd

To Ghaziabad

Siwal Khas

Dhauri

Aminagar Sarai

E. Yamuna Canal

Baraut

Yamuna R.

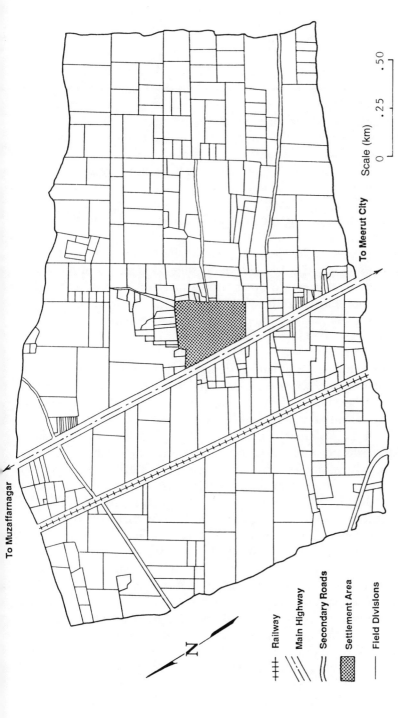

To Muzaffarnagar

To Meerut City

Scale (km)

0   .25   .50

N

┼┼┼┼ Railway

Main Highway

Secondary Roads

Settlement Area

Field Divisions

Source: Meerut, District Land Records Office, Collectorate.

Figure 3.2. Walidpur village, Meerut District

Profile of Walidpur Village

Walidpur is located in Daurala Block of Meerut District (Figs. 3.1 and 3.2). It lies on the main Meerut-Muzaffarnagar highway, about 18 kilometers from the city of Meerut and about 3 kilometers from the block headquarters town of Daurala. With a population of about 2,000, Walidpur is a medium-sized village and looks much like other villages in the area. The settlement area, or *abadi*, has both *pucca* (brick) and *kutcha* (mud) houses, and some in the process of transformation from *kutcha* to *pucca*. The lanes between houses are narrow, but many have been paved with bricks; those which are not become quite impassable during the rainy season when carts and cattle churn up the mud.

The village land lies on either side of the main highway and the railway line (Fig. 3.2). All fields are within 2 kilometers of the settlement area, so farmers do not have to travel very far to reach their plots of land. Plots vary between 4.5 hectares and less than 0.1 hectare. Consolidation operations have been conducted in Walidpur twice since the abolition of *zamindari*: first between 1955 and 1960 and a second time between 1979 and 1984. As a result, landholdings are quite compact. Small and marginal cultivators generally have one consolidated plot; when there are two or more fragments, these are close together. Some of the big farmers, however, cultivate land at two or three different locations.

The soil in Walidpur is the fertile alluvium of the Gangetic plain. The groundwater table varies between 4 and 6 meters below ground level. Almost all households draw their supply of drinking water from hand pumps rather than open wells, which was the system in the mid-1960s. The fields are dotted with tubewells, most of which run on electricity. The village was electrified in the early 1960s, and the first state tubewell was installed in 1963–64; its importance, however, has long since been diminished by private tubewells. The village is also served by minor canals from the Upper Ganga Canal. Walidpur benefited from the improved irrigation facilities of the Ganga Canal, which was opened in the mid-nineteenth century, and by the early 1900s cropping patterns had shifted in favor of sugarcane. Walidpur was in the heartland of the region that experienced the first agricultural revolution which resulted from opening the canals. In response to increased sugarcane production, a private sugar factory was established at Daurala in 1932.

The nearest railway station is at Daurala, 3 kilometers away, although a bus stop is located within the village. The residents of Walidpur complain that the government buses rarely stop at the village bus stop, and they must either walk or hire the local cycle-rickshaw to reach Daurala so that they can avail themselves of public transport services.

Comparison with Meerut District and Western Uttar Pradesh

Although Walidpur was selected for study because of the benchmark study, it proved to be a fairly representative village of Meerut District. Table 3.1 presents a comparison of some economic indicators for western U.P., Meerut, and Walidpur. Although no village can be exactly representative of the district, Walidpur has many aspects in common with the district average. The proportion of workers in agriculture in Walidpur and Meerut is almost identical—58 percent and 56 percent, respectively—indicating that the extent of off-farm diversification in Walidpur is almost on a par with the district average. The distribution of small and marginal landholdings, irrigated area, and cropping intensities are also close to the district norms.

Table 3.1. Comparison of selected economic indicators for western Uttar Pradesh, Meerut District, and Walidpur

Indicator	Western U.P.	Meerut District	Walidpur village
Agricultural workers as percentage of total work force, 1981	69	56	58
Small and marginal landholdings as percentage of total landholdings, 1980–81	82	78	78
Average area per landholding (ha), 1980–81	1.22	1.33	0.73
Net irrigated area as percentage of net cropped area, 1985–86	77	93	100
Cropping intensity (percent), counting surgarcane as double crop, 1985–86	169	214	219
Sugarcane area as percentage of gross cropped area, 1985–86	13	31	43
Wheat area as percentage of gross cropped area, 1985–86	35	33	28
Yield of sugarcane (tons/ha), 1985–86	50.9	47.8	55.5[a]
Yield of wheat (tons/ha), 1985–86	2.5	3.3	3.7[a]
Yield of rice (tons/ha), 1985–86	1.8	1.8	3.4[a]

Sources: Uttar Pradesh, Department of Agriculture, Directorate of Agricultural Statistics and Crop Insurance, *Uttar Pradesh Ke Krishi Ankre* [Agricultural statistics of Uttar Pradesh] (Lucknow, various years); Meerut, District Land Records Office, Collectorate.
[a]Yields are for 1988–89.

The average landholding in Walidpur according to the 1980–81 census was 0.73 hectare, almost half the size of the average landholding in Meerut. The distribution of land in Walidpur is extremely unequal. In 1958–59 more than 70 percent of households were landless. Of the landowning households, the 20 percent in the medium and big landholding categories owned 93 percent of the land. It is likely that to escape from the consequences of the Land Ceiling Acts of the 1960s, the land in the possession of big landowners was subdivided into smaller holdings, which were spared by the Ceiling Act. This would explain the large number of holdings in Walidpur and consequently the very small size of the average operational area per holding. If households instead of landholdings are considered, the average amount of land owned per household was about 2.1 hectares in 1983–84 (Tyagi 1988).

The sugarcane area in Walidpur, at 43 percent of the gross cropped area, is higher than the district average. The location of a sugar factory about 3 kilometers from the village, the excellent irrigation facilities, and the increasing diversification of the village economy, which is leading to labor shortages for agricultural operations, account for the comparatively high proportion of area under sugarcane.

Yields of wheat and sugarcane, at 3.7 and 55.5 tons per hectare, respectively, are higher than the district average. Rice yields in Walidpur are almost twice the district average. The many private tubewells, which ensure timely irrigation, account for the impressive performance of rice varieties in Walidpur.

Table 3.1 shows that while Walidpur is fairly representative of Meerut District, it appears more developed than the average western U.P. village. The same applies to Meerut District when compared with the western region. This, however, suits the purpose of our study. Areas that are agriculturally the most developed are the ones where diversification of the rural economy is most likely to show up.

Walidpur and Daurala Growth Center

Daurala, a small town, had a population of more than 9,000 in the 1981 census. Walidpur is significantly influenced by the Daurala growth center, and in turn, the economy of Daurala is fueled by agricultural activities in the cluster of villages that surround it. The present town of Daurala assumed importance when a private sugar mill was opened in 1932, at which time it was no more than a small village. In the 1950s, after the town had developed some infrastructure as a result of agroindustry, the government decided to locate the headquarters of the Block Development Officer (BDO) at Daurala. This brought in its wake additional infrastructure and essential services.

Today Daurala is an important terminus for government road transport; it has a railway station, government health care centers, educational institutions, distribution outlets for agricultural inputs, veterinary facilities, bank and post and telegraph offices. Moreover, all government development activities pertaining to Daurala Block are conducted at the BDO's office at Daurala. In addition to the services provided by the government, a private economy thrives in Daurala. It started in 1932 with the establishment of the Daurala Sugar Works. Since then, several bakeries, dealers in wood and wood products, chemical plants, and electrical workshops have sprung up. There is also a large number of distributors of agricultural inputs such as fertilizer, seeds, and chemicals, agricultural machinery pumpsets, tubewells, threshers, implements, and agroservice centers such as garages and welding and repair shops. Supporting the economic activities is a number of service establishments, private doctors, drugstores, tailors, barbers, tea stalls and eating houses, cloth merchants, and vendors of radios, televisions, and other consumer goods.

Walidpur does not exhibit any significant emigration characteristics. Those who leave the village do so on account of jobs in the police, army, or other government employment. There was scarcely any recollection of a household or individual emigrating because of economic distress. A few residents work in Meerut City, but their families continue to reside in the village and they visit them every weekend or fortnight.

The Benchmark and Present Studies

The AERC benchmark study of 1963–64 allows one to compare several parameters of demographic and household characteristics, agrarian structure, and agricultural performance, a comparison that would not have been possible on the basis of village land records alone. It also permits a comparison of cropping patterns by size of landholding group, which provides an indication of how the various categories of farmers have responded to changes brought about by the second-generation effects of the Green Revolution technology.

The AERC study of Walidpur is an excellent and very detailed piece of work. It suffers from one limitation, however: household income is estimated only for landowning households, referred to in the study as cultivating households; landless households are excluded. Considering the fact that the latter constituted over 70 percent of the households then in the village, excluding this category eliminated a large proportion of Walidpur, especially the poor households.

The present study, which focuses on income diffusion among the poorer rural households, takes a more comprehensive view of the village and includes landless households. This ensures that the poor households are adequately represented. The two studies are therefore not strictly comparable.

Data Collection

Data for Walidpur were collected during the 1988–89 agricultural year. There were three stages in the data collection process: a preliminary village survey, a general village survey, and a household sample survey.[1]

The starting point of the preliminary village survey was the 1981 census. Table 3.2 shows population trends since 1961. The population increased by about 2 percent annually during the 1960s and 1970s, and the average household size exhibited steady growth. Not surprisingly the amount of land per person fell from 0.24 to 0.16 hectare.

The preliminary village survey was conducted in May 1988 to assess the total number of households and the landownership pattern. This information served as the basis for the sampling procedure through which households were selected for the study. The total num-

Table 3.2. Demographic indicators for Walidpur, 1961, 1971, 1981

	1961	1971	1981
Population	1,089	1,345	1,620
Total households	198	211	262
Average size of household	5.5	6.4	6.2
Land area per person (ha)	0.24	0.18	0.16

Source: India, Office of the Registrar General, *Census of India*, General Population Tables, Series 1 (Delhi, various years).

1. The detailed methodology and data collection process employed in these surveys are discussed in Chapter 4.

Table 3.3. Total households and composition of sample for Walidpur, 1988–1989

| Households | Village total | | Sample size |
	Number	Percentage	
Landless	200	62	40
Landowning	120	37	24
Marginal/near-landless (<1 ha)	70	21	14
Small (1–2 ha)	20	6	4
Medium (2–5 ha)	29	9	6
Big (>5 ha)	1	0	—
TOTAL	320	100	64

Source: Walidpur preliminary village survey, 1988–89.

ber of households and the composition of the sample for the village are shown in Table 3.3. The households were divided into five categories by amount of land owned: landless, marginal/near-landless (<1 ha), small (1–2 ha), medium (2–5 ha), and big (>5 ha). A stratified random sample of sixty-four, 20 percent of the village household population, was selected for the study. From each group households were selected in proportion to their respective populations.

The general village survey was conducted to provide background information on the village. The survey involved interviews with a cross-section of the village population as well as with government officials at the block and district levels. The household sample survey involved administering a pretested questionnaire to the sixty-four households. The questionnaire was administered twice: once during the *kharif* and then again during the *rabi* season of the 1988–89 crop year. The information from the household sample survey was used to estimate household income from various sources.

Impact of the Green Revolution

Between the pre–Green Revolution period and the late 1980s there were significant changes in the agrarian structure and the production system in Walidpur. Among the more visible changes was in the nature of irrigation, which was transformed from largely well and canal irrigation to private tubewells. As to cropping patterns, there was a marked increase in the cultivation of wheat and sugarcane and a decline in the acreage of coarse cereals, pulses, and oilseeds. In the later years the emergence of fruit orchards marked a new trend in cash crops. Wheat and rice yields more than doubled, although no similar improvement was observed in sugarcane. Input use changed, with greater emphasis on purchased inputs and mechanization by the bigger farmers. The importance of draft cattle declined as agricultural machinery began to play a greater role in operations. Cropping intensities increased.

Changes in Agrarian Structure

The principal changes in agrarian structure have been the declining size of land held per household and the worsening of land distribution. These are largely the consequences of demographic pressures. There is no evidence of big farmers buying out small, uneconomic holdings, as was predicted by the critics of the Green Revolution. If anything, some small landholders purchased land from big farmers in the adjoining villages. The positive impact of the second-generation effects of agricultural technology is visible in the increasing proportion of workers in nonagricultural activities. An interesting trend that seems to be emerging is the increase in the incidence of tenancy and its changing nature.

Household Profile and Size of Farm. The comparative household profile of Walidpur in the pre– and post–Green Revolution periods is shown in Table 3.4. A household is defined as a group of persons who live together and share a common kitchen. In 1963–64 Walidpur had 201 households, which had increased to 320 by 1988–89. This implies an increase of about 60 percent in twenty-five years.

Table 3.4. Comparative profile of Walidpur households, 1963–1964 and 1988–1989

	1963–64	1988–89
Total households	201	320
Number of landless households	143	200
(percentage of total households)	(71.1)	(62.5)
Number of landowning households	58	120
(percentage of total households)	(28.9)	(37.5)
Total operational area (ha)	212	197
Operational area per landowning household (ha)	3.65	1.64

Sources: S. S. Tyagi, "Walidpur: Agricultural Transformation in Two Decades." Research Study 88/3, Agro Economic Research Centre, Delhi University, 1988; Walidpur preliminary village survey, 1988–89.

While in absolute terms the number of landless households has risen, their percentage of the total has declined. The largest percentage increase has been among the marginal households. The increase in the number of households has come about largely as a result of subdivision of households. Land reform measures undertaken in the mid-1970s, when surplus village common land was distributed in small parcels to the landless households of the village, added about twenty-five marginal households. Most of these holdings are smaller than 0.1 hectare (Fig. 3.2).

Government antipoverty programs have also resulted in an increase in the number of landless and marginal households. Households should fall into the landless or marginal categories to qualify for assistance under these programs; further, the benefits per household are usually capped at Rs 3,000. Therefore, if each adult member in a family is to avail himself or herself of the maximum benefit, each must maintain a separate household. Considering medium and big households together, their number has declined from thirty-nine to thirty, while small households have increased from nine to twenty.

Table 3.4 also indicates that the size of operational area per landowning household has declined from an average of 3.7 to 1.6 hectares. While the primary reason for this decrease is the increase in the number of households, consolidation operations, which were conducted during 1979–85, also resulted in some decline in the total operational area, with some land diverted to nonagricultural uses.

Land Distribution. Figure 3.3 indicates distribution of households and the area they operate by landholding category. Note that landless households are included. In 1963–64 landless households constituted more than 70 percent of the total. In 1988–89 this number had declined to 63 percent. Among landholding households the principal changes are the virtual disappearance of the big landholder category as the result of demographic pressure, the emergence of medium households as the group controlling the bulk of the land, and the growing importance of small and marginal operators as cultivators of the land. Despite these changes, however, a comparison of the Gini coefficient for the two periods indicates a slight worsening of land distribution in Walidpur.

Tenancy: Increasing Trend in Sharecropping. The benchmark study in 1963–64 reported that tenancy played an insignificant role in Walidpur, accounting for a mere 1 percent of the total area under

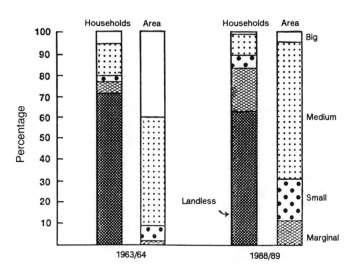

Source: Table A3.1.

Figure 3.3. Percentage distribution of households and land owned in Walidpur, by landholding group, 1963–1964 and 1988–1989

cultivation. This low figure probably reflects the fact that tenancy is very difficult to document. There are no official records of tenancy, and contracts between landlord and tenant are verbal. Farmers are very reluctant to divulge information on the subject. Hence it is likely that the figure of 1 percent reported in 1963–64 is an underestimation. In 1988–89 half of the forty landless households in the household sample survey reported that they were involved in sharecropping and that total sharecropped area was about 15 percent of net cultivated area.

The benchmark study noted that leasing of land in 1963–64 although very limited, was generally arranged on a yearly basis. The nature of tenancy seems to have undergone a major change since then. In 1988–89 all tenancy contracts in Walidpur were on a sharecropping basis. Contracts are made not on a yearly arrangement but rather by crop. From the landowner's point of view the motivation for sharecropping appears to be the difficulty in procuring labor at peak agricultural seasons. With multiple cropping practices intensifying the need for timely harvest and field preparation, and village workers now having employment opportunities outside agriculture, farmers often face acute labor shortages during peak operations. A sharecropping arrangement with a landless or near-landless household is beneficial for the landlord as it reduces the cost of searching for and supervising labor.

Landless households also stand to benefit from sharecropping. In addition to providing access to land, sharecropping also serves as a cost- and risk-sharing mechanism for such crops as potatoes and vegetables. Furthermore, it allows the sharecropper to cut grass, weeds, and sugarcane tops for his livestock. In ordinary circumstances, landowners' fields are out-of-bounds to landless households.

It is not surprising that the bulk of sharecropping in Walidpur is for paddy, potatoes, and vegetables—all labor-intensive crops. There is no sharecropping for wheat and sugarcane, which are relatively less labor-intensive and for which most operations, except harvesting, are becoming mechanized.

Composition of Village Work Force. Figure 3.4 shows the number of workers in agricultural and nonagricultural activities as reported by the census. Data from the 1961 census are not included; they are not comparable to 1971 and 1981 data because of major changes in

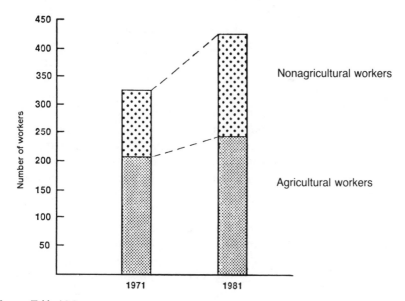

Source: Table A3.2.
Note: Nonagricultural includes animal husbandry, household industry, manufacturing, construction, trade, transport, and services; agricultural includes cultivators and agricultural laborers.

Figure 3.4. Composition of Walidpur work force, 1971 and 1981

the definitions of some categories. According to the census definition, agricultural workers include cultivators and agricultural laborers. Nonagricultural workers are those engaged in household industry, manufacturing, construction, services and repairs, trade, commerce, business, mining, and the like. In 1971 the total number of workers in Walidpur was 329, of whom 37 percent derived the major share of their income from nonagricultural activities. In 1981 the total number of workers increased to 423, and the proportion of workers in non-agricultural activities increased to 42 percent. No official census figures exist for 1988–89, however, the household sample survey indicates that at least 60 percent of households in the village have nonagricultural activities as their main occupation, and 20 percent participate toward nonagricultural pursuits on a part-time basis. It is clear that there is an increasing trend toward nonagricultural sources of income.

Changes in Agricultural Performance

The impact of Green Revolution technology on agricultural performance in Walidpur can be observed in the changing nature of irrigation, cropping patterns, and yields of wheat and rice. The early period, from the mid-1960s to the mid-1970s, witnessed the direct impact of technical change in terms of dramatic shifts in cultivated area in favor of wheat as well as an impressive increase in wheat yields. By the late 1980s, the second-generation effects of the Green Revolution, which were more prominent in their stimulation of rural off-farm economy, began to indirectly influence agricultural performance. In particular, labor shortages prompted big farmers to shift to less labor-intensive crops such as sugarcane and fruit orchards. Already noted is the rising incidence of sharecropping for labor-intensive crops.

Increasing Importance of Tubewell Irrigation. The nature of irrigation has undergone major changes since the mid-1960s. Figure 3.5 indicates changes in gross cropped area, gross irrigated area, and area irrigated by tubewells. Whereas only 20 percent of the irrigated area was served by tubewells in 1963–64, more than 85 percent was so watered in the mid-1980s. Without the assured and controlled irrigation made possible by tubewells, the spread of modern varieties of wheat and rice could not have taken place.

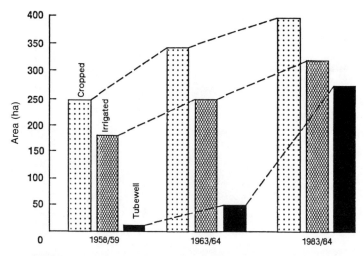

Source: Table A3.3.

Figure 3.5. Gross cropped area, gross irrigated area, and tubewell irrigated area in Walidpur, 1958–1959, 1963–1964, and 1983–1984

Cropping Intensity and Cropping Patterns. Figure 3.6 shows the cropping intensity in Walidpur from 1958–59 to 1988–89. The cropping intensity has risen consistently over the past three decades, reflecting major changes in cropping patterns.[2]

Figure 3.7 illustrates the changing cropping pattern in Walidpur village between 1958–59 and 1988–89. The initial impact of the Green Revolution is seen in the impressive increase in the area planted to wheat between 1963–64 and 1970–71. The other striking early consequence was the reduction in the area under pulses and oilseeds, which declined by almost three-fourths between the mid-1960s and early 1970s. Paddy, which was negligible in the pre–Green Revolution period, seems to be increasing.

Toward the late 1980s some new trends began to appear. Pulses and oilseeds, which were initially adversely affected, began to reappear but in a different cropping season. Most of the pulses and oilseeds that suffered a setback were *rabi* crops—gram and mustard—which could not compete with HYVs of wheat. However, with the development of short-duration varieties of redgram and greengram and rapeseed, which do not compete with wheat but can be harvested

2. Cropping intensity has been estimated by treating sugarcane as a double crop.

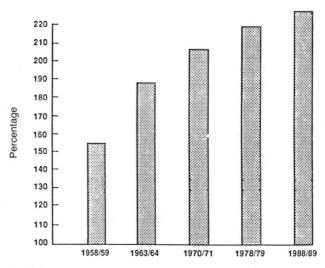

Source: Table A3.4.

Figure 3.6. Cropping intensity in Walidpur, 1958–1959, 1963–1964, 1970–1971, 1978–1979, and 1988–1989

as bonus sixty-day crops, the acreage under pulses and oilseeds seems to be picking up. Another recent trend that exhibits considerable potential is the increase in fruit and vegetable acreage. In Walidpur it is largely mango orchards that are gaining ground.

The consistent increase in sugarcane area despite no major breakthrough in yields, and also the fact that the crop occupies the land throughout the year, is an indication that medium and big farmers are attracted to this crop as a means of dealing with labor shortages and increasing wage rates of agricultural labor. The increased area under fruit orchards is a reflection of the same trend.

Cropping Patterns by Farm Size. An examination of cropping patterns by farm size provides a better understanding of the impact of second-generation effects on cropping patterns. Figure 3.8 shows the net and gross cropped area per household by landholding group as well as the percentage distribution of major crops in the gross cropped area by landholding group for 1963–64 and 1988–89. The net and gross cropped areas are read on the right-hand scale of the figure, while the percentage share of the various crops is read on the left-hand scale.

The figure indicates that for the marginal/near-landless households

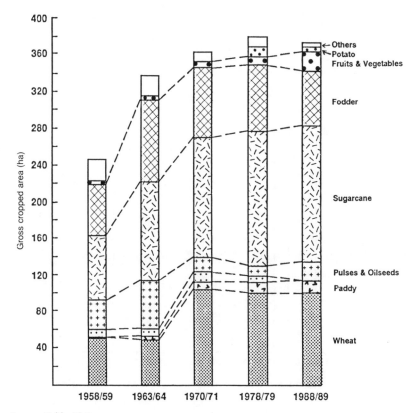

Source: Table A3.5.

Figure 3.7. Gross cropped area under major crops in Walidpur, 1958–1959, 1963–1964, 1970–1971, 1978–1979, and 1988–1989

the net cultivated area is about 0.3 hectare, while gross cropped area is twice that size. The major change between the two periods occurred in wheat acreage, which increased from 17 to nearly 40 percent of the gross cropped area. Paddy, which was not cultivated at all in the earlier period, seems to have replaced coarse cereals during the *kharif* season. Sugarcane acreage has been reduced considerably by marginal cultivators, who prefer to engage in multiple cropping rather than occupy the land for a whole year with sugarcane. The area under fodder registered a major increase as dairying acquired greater importance among marginal households.

The average household in the small landowner category cultivated 1.7 hectares net and about 3 hectares gross. The principal change between the two periods for this group was the substitution of wheat

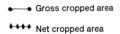

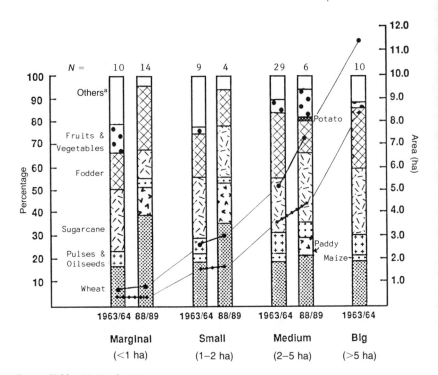

Source: Tables A3.6 and A3.7.
[a] Others: coarse cereals, spices, cotton.

Figure 3.8. Cropping patterns, by landholding group, in Walidpur, 1963–1964 and 1988–1989

for *rabi* pulses and coarse cereals and the emergence of paddy as a major *kharif* crop.

Cultivators in the medium category increased their wheat acreage relatively little. Instead they expanded their paddy and sugarcane production. This is the only category in which the proportion of land in sugarcane increased. Orchard fruits have also emerged as important cash crops for medium-level farmers. Fodder, on the other hand, has lost its earlier importance. This is in keeping with the trend of increased mechanization, which has considerably reduced the need for draft animals. The reduction in draft animals has not been offset by increased numbers of milk cattle, hence the overall reduction in fodder acreage.

Yields of Major Crops. Figure 3.9 shows the yield of major crops in the two time periods. It is clear that wheat and paddy yields have increased impressively. There has not been any significant change in sugarcane yields.

Income Diffusion over Time

The direct impact of the Green Revolution in Walidpur is most clearly evident in the increased acreage given to HYVs of wheat and rice, especially by small and marginal cultivators, which has resulted

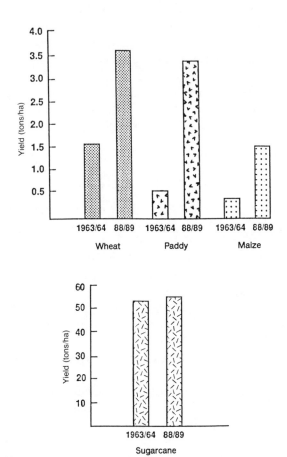

Source: Table A3.8.

Figure 3.9. Yields of major crops for Walidpur, 1963–1964 and 1988–1989

in higher cropping intensities and has changed the nature of irrigation in the village. The secondary effects of the Green Revolution technology are also visible in the changes in cropping patterns in the late 1980s. While the small and marginal cultivators have increased wheat, rice, and fodder acreage, the medium farmers have increased their acreage of sugarcane and orchards and decreased their fodder area. This implies a trend toward increasing mechanization and a preference for less labor-intensive crops, reflecting a shortage of labor for agricultural operations during peak demand periods since much of the village's labor has been diverted into nonagricultural activities.

The benchmark study provides some indication of the level and distribution of household income in 1963–64. Unfortunately there is no mid-1970s study to provide an indication of income changes in the early phase of the Green Revolution, when the direct impact of the new technology was more pronounced and the secondary effects had not begun to significantly influence incomes of the rural poor. By 1988–89, the direct impact, as reflected in crop and agricultural labor incomes, had become somewhat diluted by the secondary effects, captured mainly in the off-farm earnings of the landless and near-landless households.

Changes in the Level of Household Income

Figure 3.10 shows income estimates for the sample households in Walidpur by size of landholding group for the crop years 1963–64 and 1988–89. The household and per capita incomes, in constant 1988–89 prices, are read on the right-hand scale of the figure. The left-hand scale indicates the percentage contribution of each income source to the mean household income of the particular category of landholders. For instance, the real annual household income of the marginal landholding group increased from about Rs 9,000 in 1963–64 to more than Rs 16,000 in 1988–89, while the contribution of crop income to these totals fell from greater than 40 percent to less than 35 percent.

It appears that household income has risen impressively in all three of the landholding categories for which a time comparison is possible: by 80 percent in the marginal group, 73 percent in the small, and 106 percent in households in the medium category. The 1963–64 study did not include landless households, and the 1988–89 sample did not include big landowner households, as they had by then been reduced to an insignificant number. Thus it is not possible to compare these two categories over time.

These changed income levels may or may not accurately depict the evolution of Walidpur's economy. The rate of income growth is heavily influenced by the cost-of-living index used to convert the 1963–64 figures into 1988–89 rupees. The only available index specific to Uttar Pradesh is the Consumer Price Index for Agricultural Laborers, which does not necessarily apply to the broad spectrum of households in Walidpur. That it may not is suggested by the growth rates in per capita incomes that emerge: 2.38 percent for the marginal households, 1.52 percent for the small, and 1.36 percent for households in the medium group. The latter two differ little from the 1.5 percent rate believed to apply to India's per capita gross domestic product during the years since independence; given the extent of change in Walidpur, higher rates would have been expected. Perhaps

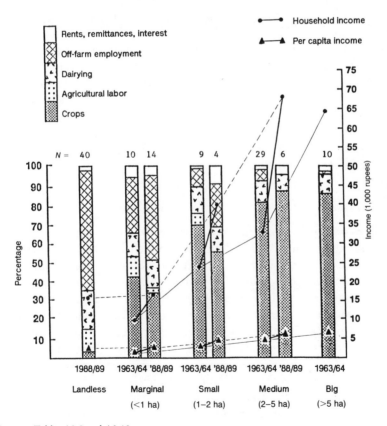

Sources: Tables A3.9 and A3.10.

Figure 3.10. Amount and sources of income for Walidpur, by landholding group, 1963–1964 and 1988–1989 (income in 1988–1989 prices)

a better indicator of the changes that have taken place in Walidpur is the incidence of poverty.

Declining Incidence of Poverty. The concept of the poverty line is used by various official agencies to estimate the proportion of the population living in poverty and as a criterion for selecting beneficiaries for government antipoverty programs. As defined by the Planning Commission for use in rural India, it refers to the level of expenditure needed by a household to purchase 2,400 calories of food for each of its members per day. The poverty line for a household of five members was estimated at Rs 6,400 in 1984–85 prices (India, Planning Commission 1985). Taking account of the increase in the cost-of-living index for agricultural laborers, the poverty line in 1988–89 prices was about Rs 8,600 per household and Rs 1,700 per capita.

The benchmark study reported that in 1963–64 six of the fifty-eight households, or about 12 percent of all cultivating households, had annual incomes below the poverty line. Of these, a majority were in the marginal/near-landless category. In 1988–89, six of the sixty-four sample households, or about 9 percent, had annual incomes below the poverty line. Of these, four were from the landless category and two were from the marginal category. Since the 1988–89 sample included landless households and the 1963–64 sample did not, it seems reasonable to conclude that there has been a significant decline in the incidence of poverty in Walidpur.

Changes in the Composition of Household Income

The composition of household income by landholding group is also illustrated in Figure 3.10 and may be a better indicator of the changes that have occurred since 1963–64 than the figures for income. Four income sources predominate: crop cultivation, wages for agricultural labor, dairying, and off-farm employment. To these revenues are added modest amounts from such miscellaneous sources as remittances from family members living away from the village, rents, pensions, and interest on savings. It is not surprising that income from crops is most important to households in the small, medium, and big landholding categories, and noncrop sources contribute the greater part of the revenues of landless and marginal households. The growing importance of off-farm employment as an income source is especially striking.

For the landless category more than 95 percent of income in 1988–

89 was derived from noncrop sources. Off-farm employment contributed an impressive 64 percent of this, followed by dairying (20 percent), agricultural labor (12 percent), and rents and remittances (2 percent). The share of crop income, although negligible, was not zero, because landless households are frequently involved in crop cultivation through sharecropping contracts, usually for one crop season.

In marginal households, crop income constituted over 40 percent of total income in 1963–64; this declined to less than 35 percent in 1988–89. Agricultural labor income also declined significantly, while nonfarm activities gained in relative importance. Off-farm income sources were the major gainers, and their share in total income rose by one and a half times. The proportion of income from dairying rose marginally, and that from rents and remittances declined.

The contribution of crop income declined from around 70 percent to 55 percent of the total in the case of small landholder households. The element of agricultural labor almost disappeared for this category. Of the noncrop sources of income, off-farm employment increased significantly, as did earnings from rents and remittances.

The medium-level landholding households, unlike the other categories, increased their share of farm income from crop cultivation. Off-farm activities, which contributed a small proportion to total income in the earlier period, virtually disappeared in 1988–89; however, the share of rents and remittances increased marginally.

A comparison of the composition of total household income by source and by size of landholding group indicates that the noncrop sector has gained in importance in all but the medium-level landholding households. The contribution of off-farm earnings to the income of landless and marginal households is substantial and is likely to increase in the future. Although crop cultivation continues to be the major occupation in small farm households, diversification of income sources is occurring, with off-farm earnings increasing their share in household income.

Changes in Income Distribution

The best single measure of inequality is the Gini coefficient of concentration, defined as the ratio of the area between the Lorenz curve and the diagonal, and the total area under the diagonal. A Lorenz curve is a plot of the cumulative proportion of units, arrayed in order from the smallest incomes to the largest, against the cumulative share of the aggregate income accounted for by these units. Perfect equality would result in points along the diagonal, and in cases of perfect inequality points would be along the baseline and the right-hand vertical axis. Actual curves fall in between, and the closer to the diagonal, the less the inequality. The Gini coefficient has a maximum value of unity (absolute inequality) and a minimum value of zero (absolute equality).

Figure 3.11 shows the Lorenz curves and Gini coefficients for income distribution for 1963–64 and 1988–89. The Lorenz curves and the Gini ratios must be interpreted in light of the fact that the 1963–64 study included only landowning households. The distribution of income in 1963–64 is therefore between marginal households at the one extreme and big landowner households at the other. The 1988–89 study sample included landless households in proportion to their presence in the village, and therefore a wider—and poorer—spectrum of the population was represented. The landless households are at one end while medium landholder households are at the other. There were no big landholders in the sample.

Despite these problems of incomparability, some useful insights can be drawn from the Lorenz curves of the two income distributions. The Gini coefficients in 1963–64 and 1988–89 were 0.26 and 0.25, respectively. Since the Gini coefficient in the second period remained almost the same despite the inclusion of a larger proportion of poorer households, it is reasonable to conclude that income distribution in 1988–89 was more equitable than in the earlier period.

Anticipating Future Income Distribution

The salutary effect of noncrop income on the distribution of total income is especially obvious in 1988–89, when, because of the inclusion of so many landless households, the distribution of crop income was particularly skewed. For purposes of policy intervention it would be desirable to know more about the impact of the several sources of noncrop income. A technique has been developed to split the Gini coefficient into its constituent parts, making this possible. This technique is known as the decomposition of the Gini coefficient.[3] One can also now estimate the elasticity of total inequality by income source. These elasticities provide a means for anticipating changes in total income inequality likely to result from future changes in income from

3. Discussion of the decomposition of the Gini coefficient is provided in the appendix at the back of the book.

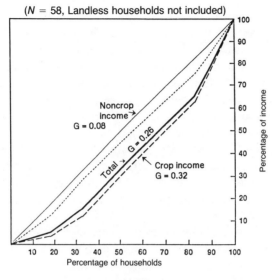

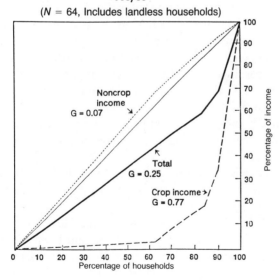

Source: Table A3.11.

Figure 3.11. Lorenz curves for Walidpur, by landholding group, 1963–1964 and 1988–1989

the several sources (Boisvert and Ranney 1990, Lerman and Yitzhaki 1985, Pyatt et al. 1980).

Table 3.5 presents the results of applying the technique to the findings of the 1988–89 Walidpur survey. The inflated Gini coefficient for crop income ($G = 0.77$) is not surprising in view of the large proportion of landless households in the sample. The disparity in crop income is primarily a function of the unequal landownership.

The Gini coefficient for dairying ($G = 0.11$) points to a more equitable distribution of income from this source than from crops. Dairy enterprises are as widespread among the landless and near-landless as among larger landholders. The demand for milk and dairy products, the improvement in rural infrastructural facilities, and government intervention to assist the rural poor in acquiring milk animals have all contributed to the rapid spread of dairying among households in the landless and near-landless categories. The Gini coefficient for off-farm employment ($G = 0.21$) suggests that this income source also has a beneficial impact on the distribution of income.

Two other factors in combination with the Gini coefficient show the actual share of inequality: the proportion of income of the particular source, S, and the location of recipients of income from this source in the total income distribution, R. It is the combined impact of S, G, and R (column 5) that reflects the importance of the individual income sources in influencing overall inequality ($G = 0.25$). It is evident that crop income is the most significant source of inequality and that dairying contributes the least to inequality. Agricultural labor and off-farm income have a counterbalancing effect, with off-farm revenues being the more important of the two. This offsetting influence derives from the negative correlation between agricultural labor and off-farm employment (column 4) and total income. This is expected, since it is chiefly the landless and the marginal households that hire themselves out as farm laborers or engage in off-farm employment.

To predict the likely impact on income distribution of future changes in income from any one source, one must consider the elasticity of total inequality by income source. These elasticities (column 7) suggest that any economic change that brings about a 1 percent increase in crop income, ceteris paribus, will exacerbate inequality by almost 0.80 percent, whereas an equal increase in income from off-farm employment will reduce inequality by 0.61 percent. Increases in agricultural labor and dairying income would also reduce inequality—by 0.13 and 0.11 percent, respectively.

Table 3.5. Decomposition of income inequality by income source in
Walidpur, 1988–1989

Income source	Income share (S)	Gini of source (G)	Correlation with rank of total income (R)	Share of inequality		Elasticity of total inequality by income source
				SGR	Percentage	
(1)	(2)	(3)	(4)	(5)	(6)	(7)
Crop income	0.37	0.77	1.00	0.29	116.9	0.80
Agricultural labor	0.06	0.45	−0.69	−0.02	−7.2	−0.13
Dairying	0.15	0.11	0.56	0.01	3.7	−0.11
Off-farm employment	0.39	0.21	−0.68	−0.05	−22.1	−0.61
Rents, remittances, pensions, interest	0.04	0.57	1.00	0.02	8.7	0.05
TOTAL	1.00	0.25	1.00	0.25	100.0	—

Source: Walidpur household sample survey, 1988–89.

The elasticities of total inequality suggest that changes in both crop
and off-farm income can significantly affect income distribution. The
latter has the potential to reduce income inequality. Thus, rural development efforts that promote greater diversification into off-farm activities are most likely to reduce income inequality. Increases in dairying income and labor-intensive cropping patterns would also have a
salutary effect on income distribution.

Appendix

Table A3.1. Percentage distribution of households and land owned in Walidpur, by landholding group, 1963–1964 and 1988–1989

Size group	1963–64 Households	Area	1988–89 Households	Area
Landless	71.0	0	62.5	0
Marginal	5.0	1.6	21.9	12.3
Small	4.5	6.9	6.3	18.4
Medium	14.0	50.2	9.1	65.8
Big	5.4	41.3	0.3	3.5
TOTAL	100	100	100	100
Gini coefficient		0.81		0.82

Sources: S. S. Tyagi, "Walidpur: Agricultural Transformation in Two Decades." Research Study 88/3, Agro Economic Research Centre, Delhi University, 1988; Preliminary village survey, 1988–89.

Table A3.2. Composition of Walidpur work force, 1971 and 1981 (number of workers)

Workers	1971	1981
Agricultural	206	246
Nonagricultural	123	177
TOTAL	329	423

Source: India, Office of the Registrar General, *Census of India*, General Population Tables, Series 1 (Delhi, various years).

Note: Agricultural includes cultivators and agricultural laborers; nonagricultural includes animal husbandry, household industry, manufacturing, construction, trade, transport, and services.

Table A3.3. Gross cropped, gross irrigated, and tubewell irrigated area in Walidpur, 1958–1959, 1963–1964, and 1983–1984 (ha)

	1958–59	1963–64	1983–84
Gross cropped area	248.0	341.2	385.7
Gross irrigated area	183.7	249.7	320.5
Area irrigated by tubewells	10.1	49.4	276.0

Sources: R. R. Vaish, *Walidpur.* Agro Economic Research Centre Report 54 (Delhi: Delhi University, 1964); S. S. Tyagi, "Walidpur: Agricultural Transformation in Two Decades." Research Study 88/3, Agro Economic Research Centre, Delhi University, 1988.

Table A3.4. Cropping intensity in Walidpur, 1958–1959 through 1988–1989

	1958–59	1963–64	1970–71	1978–79	1988–89
Net cropped area (ha)	209.4	238.0	239.2	239.2	227.4
Gross cropped[a] area (ha)	324.6	449.8	497.5	529.7	523.3
Cropping intensity (percentage)	155	189	208	221	230

Sources: R. R. Vaish, *Walidpur.* Agro Economic Research Centre Report 54 (Delhi: Delhi University, 1964); S. S. Tyagi, "Walidpur: Agricultural Transformation in Two Decades." Research Study 88/3, Agro Economic Research Centre, Delhi University, 1988, p. 17; Meerut, District Land Records Office, Collectorate.
[a]Sugarcane is counted as a double crop.

Table A3.5. Gross cropped area under major crops in Walidpur, 1958–1959 to 1988–1989 (ha)

Crop	1958–59	1963–64	1970–71	1978–79	1988–89
Wheat	51.6	51.0	105.2	101.2	100.7
Paddy	0.4	1.0	8.1	10.5	12.6
Maize	8.1	7.7	10.9	6.1	1.7
Pulses	25.1	51.9	15.4	10.9	13.4
Oilseeds	0.4	—	—	—	7.9
Sugarcane	75.7	108.4	131.9	147.3	149.6
Fodder	58.3	91.4	76.5	73.6	59.7
Vegetables	0.4	0.7	1.2	1.2	2.5
Fruits	2.0	—	3.6	6.9	17.7
Potato	—	2.8	1.2	11.3	5.4
Others	26.0	25.3	11.8	13.4	2.5
TOTAL	248.0	341.0	365.0	382.4	373.7

Sources: R. R. Vaish, *Walidpur.* Agro Economic Research Centre Report 54 (Delhi: Delhi University, 1964); S. S. Tyagi, "Walidpur: Agricultural Transformation in Two Decades." Research Study 88/3, Agro Economic Research Centre, Delhi University, 1988; Meerut, District Land Records Office, Collectorate.

Table A3.6. Net and gross cropped area per household in Walidpur, by landholding group, 1963–1964 and 1988–1989 (ha)

Landholding group	1963–64		1988–89	
	Net cropped area	Gross cropped area	Net cropped area	Gross cropped area
Marginal	0.33	0.60	0.32	0.62
Small	1.75	2.76	1.68	3.03
Medium	3.73	5.18	4.22	7.35
Big	8.54	11.43	—	—

Sources: S. S. Tyagi, "Walidpur: Agricultural Transformation in Two Decades." Research Study 88/3, Agro Economic Research Centre, Delhi University, 1988, pp. 80–81; Walidpur household sample survey, 1988–89.

Table A3.7. Percentage distribution of gross cropped area, by landholding group in Walidpur, 1963–1964 and 1988–1989

Landholding group	Wheat	Paddy	Maize	Pulses	Oilseeds	Sugarcane	Fodder	Potato	Fruits	Vegetables	Other
1963–64											
Marginal	16.9	—	—	6.2	—	27.6	14.7	—	—	12.4	22.2
Small	18.8	—	3.5	6.8	—	26.4	19.0	—	—	3.3	22.2
Medium	18.9	0.6	3.9	8.9	—	24.3	27.2	—	—	6.7	9.5
Big	17.5	0.4	3.7	9.1	—	29.1	25.5	—	—	3.4	11.3
1988–89											
Marginal	38.7	13.0	3.6	—	2.9	11.7	26.2	—	—	—	3.9
Small	36.0	18.0	2.0	—	—	22.0	16.0	—	—	—	6.0
Medium	22.2	7.0	—	—	8.0	30.0	14.0	1.0	12.4	—	5.4

Sources: S. S. Tyagi, "Walidpur: Agricultural Transformation in Two Decades." Research Study 88/3, Agro Economic Research Centre, Delhi University, 1988; Walidpur household sample survey, 1988–89.

Table A3.8. Yields of major crops in
Walidpur, 1963–1964 and 1988–1989
(tons/ha)

Crop	1963–64	1988–89
Wheat	1.60	3.65
Paddy	0.66	3.40
Maize	0.42	1.60
Sugarcane	53.50	55.54

Sources: S. S. Tyagi, "Walidpur: Agricultural Transformation in Two Decades." Research Study 88/3, Agro
Economic Research Centre, Delhi University, 1988; Walidpur household sample survey, 1988–89.

Table A3.9. Annual household and per capita income in Walidpur, by
landholding group, 1963–1964 and 1988–1989 (rupees)

Landholding group	Current prices		Constant prices[a]	
	1963–64	1988–89	1963–64	1988–89
Per household				
Landless	—	15,958	—	15,958
Marginal	1,369	16,291	9,044	16,291
Small	3,498	40,000	23,110	40,000
Medium	4,935	67,065	32,603	67,065
Big	9,609	—	63,482	—
Per capita				
Landless	—	2,533	—	2,533
Marginal	224	2,671	1,483	2,671
Small	437	4,211	2,890	4,211
Medium	656	6,079	4,333	6,079
Big	980	—	6,475	—

Sources: S. S. Tyagi, "Walidpur: Agricultural Transformation in Two
Decades." Research Study 88/3, Agro Economic Research Centre, Delhi
University, 1988; Walidpur household sample survey, 1988–89.
[a]Adjusted using the Consumer Price Index for Agricultural Laborers for
Uttar Pradesh; base year 1988–89 = 100.

Table A3.10. Annual household income, by source and landholding group, for Walidpur, 1963–1964 and 1988–1989 (rupees)

| Landholding group | No. of households | Crop income | Noncrop income | | | | Total household income |
			Agricultural labor	Dairying	Off-farm employment	Rents, remittances, pensions, interest	
1963–64 (N = 58)							
Marginal	10	577 (42.2)	182 (13.3)	149 (10.9)	380 (27.8)	81 (5.9)	1,369 (100)
Small	9	2,425 (69.3)	199 (5.7)	493 (14.1)	339 (9.7)	42 (1.2)	3,498 (100)
Medium	29	3,973 (80.5)	0 (0.0)	523 (10.6)	311 (6.3)	128 (2.6)	4,935 (100)
Big	10	8,201 (85.3)	0 (0.0)	1,052 (10.9)	58 (0.6)	298 (3.1)	9,609 (100)
TOTAL	58	4,010 (79.2)	63 (1.2)	563 (11.1)	290 (5.7)	140 (2.8)	5,066 (100)
1988–89 (N = 64)							
Landless	40	411 (2.6)	1,943 (12.2)	3,109 (19.5)	10,255 (64.3)	240 (1.5)	15,958 (100)
Marginal	14	5,584 (34.3)	474 (2.9)	2,358 (14.5)	7,204 (44.2)	671 (4.1)	16,291 (100)
Small	4	21,852 (54.6)	0 (0.0)	5,058 (12.6)	9,590 (24.0)	3,500 (8.8)	40,000 (100)
Medium	6	58,130 (86.7)	0 (0.0)	5,352 (8.0)	0 (0.0)	3,583 (5.3)	67,065 (100)
TOTAL	64	8,313 (37.2)	1,299 (5.8)	3,277 (14.7)	8,585 (38.5)	852 (3.8)	22,326 (100)

Sources: S. S. Tyagi, "Walidpur: Agricultural Transformation in Two Decades." Research Study 88/3, Agro Economic Research Centre, Delhi University, 1988; Walidpur household sample survey, 1988–89.
Note: Figures in parentheses are percentages of total household income.

Table A3.11. Cumulative percentage of household income in Walidpur, by source, 1963–1964 and 1988–1989

Landholding group	Households	Crop income	Noncrop income	Total income
1963–64[a]				
Marginal	17.24	2.50	12.90	4.65
Small	32.76	12.00	29.00	15.55
Medium	82.76	53.30	76.00	65.92
Big	100	100	100	100
1988–89[b]				
Marginal	62.5	3.32	69.21	44.68
Small	84.4	18.02	85.92	60.64
Medium	90.7	34.45	94.01	71.84
Big	100	100	100	100

Sources: S. S. Tyagi, "Walidpur: Agricultural Transformation in Two Decades." Research Study 88/3, Agro Economic Research Centre, Delhi University, 1988; Walidpur household sample survey, 1988–89.

[a]$N = 58$, sample does not include landless households.
[b]$N = 64$, sample includes landless households.

PHOTOGRAPHS

MEERUT DISTRICT

Meerut District in western Uttar Pradesh is in the heart of the Green Revolution belt of the Indo-Gangetic plain. The land is level as far as the eye can see and with irrigation can be cultivated throughout the year. Although densely populated, the region does not appear so, as almost all the land area is in crops (top). Mechanization is commonplace. Middle, a tractor-powered pumpset draws water from a deep bore. Almost all land preparation is done with machinery. Bottom, sugarcane in the background, wheat and mustard in the foreground.

OFF-FARM DIVERSIFICATION: RURAL GROWTH CENTERS

A key feature of change in western Uttar Pradesh during the past several decades has been the growth of small and medium-sized towns, with populations of up to 50,000 persons, that have close ties with their rural surroundings. Many such growth centers got their start because a sugar mill or similar agroindustry was located there (upper left), but increasingly they offer markets for village produce (upper right) and needed goods and services. The furniture shop (lower left) is typical of the small-scale manufacturing to be found in the growth centers, the conveyance (lower right) of the means whereby villagers find their way to town.

Off-Farm Diversification: Village-Level Activities

Typical sources of off-farm employment in the villages of Meerut District are the making of *gur*, prepared by boiling the juice of sugarcane crushed at the local *kolhu* (upper left), and handwoven cotton yard goods (upper right). The *dudhiya* has long been a feature of entrepreneurship in rural India. Twice daily he collects milk from local producers and sells it in a nearby growth center or town. Although he pays lower prices for milk than do cooperative societies, he usually does a better business because of the credit he extends to small producers. Two cans of milk are the usual load of a *dudhiya* equipped with a bicycle (lower left), six if he can afford a motorcycle (lower right).

DAIRYING AS A SOURCE OF INCOME DIFFUSION

Dairying, once important only among landowning households, has become a major income source for the landless and near-landless. Among such households, the entire dairying operation—from cutting and carrying fodder, chaffing and preparing the feed, feeding (top), cleaning and milking the animals (middle), to collecting dung and baking dung cakes (bottom)—is done by women. The government-run Integrated Rural Development Program has played an important role in this by providing credit at subsidized rates for the purchase of buffaloes.

Cultivation of High-Value, Labor-intensive Crops

The increased cultivation of such high-value crops as potatoes and vegetables is largely confined to the large and medium landholdings. Most poorer households have benefited only through the additional opportunities created for agricultural laborers. Potato cultivation remains largely manual and requires two and a half times the labor of wheat. During harvest women pick up the tubers that have been dug up by men (top) and pile them in a stack (middle). They are then graded and bagged by hand and carried to a cold-storage facility (bottom).

4

Mechanisms of Income Diffusion:
Meerut District

The analysis of income changes over time in Walidpur indicates that whereas crop cultivation is responsible for the higher incomes of farmers with holdings larger than 2 hectares, the great majority of households, which have access to less land, derive most of their earnings from noncrop activities. In Meerut District, the location of the present study, three such activities predominate. They are, in order of importance, off-farm diversification of the household economy; dairying; and the cultivation of high-value, labor-intensive crops. In the following sections we examine each of these mechanisms of income diffusion as it operates in Meerut District.

Profile of Meerut District

Meerut District, in western U.P., is the location for the present study (Fig. 3.1). Meerut, together with other districts of the Upper Ganges–Yamuna *doab*, experienced its first agricultural revolution, brought about by canal irrigation, in the mid-nineteenth century. In the mid-1960s it was part of the heartland affected by modern varieties of wheat and tubewell irrigation. By the late 1970s the second-generation effects of the Green Revolution had begun to manifest themselves in the district.

Meerut is about 65 kilometers northeast of New Delhi on an almost flat alluvial plain. The Yamuna and Ganges rivers form the western and eastern boundaries of the district, while the Eastern

Yamuna Canal, the Upper Ganga Canal, and the Anupshahar branch of the Ganges Canal run from north to south almost parallel to the two rivers. The district is divided into four subdivisions and eighteen development blocks.

Demographic Characteristics

Like other areas with fertile land and extensive possibilities for irrigation, the district is densely populated. In 1981 a population of nearly 2.8 million persons was concentrated on a land area of about 3,900 square kilometers. The population density for the district as a whole was 708 persons per square kilometer (Uttar Pradesh, State Planning Institute, *Sankhyiki Patrika, Meerut*). Rural people then accounted for 69 percent of the population, giving rural Meerut a population density of more than 500 persons per square kilometer. There were more than 900 villages and twenty-five towns in the district. There was a tremendous spurt in urban growth in Meerut District in the decade of the 1970s. Whereas the rural population grew at the annual compound rate of 1.1 percent, urban growth, at 5.6 percent, was more than five times as great. The latter was also considerably higher than the 3.1 percent growth rate experienced during the 1960s. Meerut City, the headquarters of the district, is the only Class I town; with a population of about 0.5 million, it accounts for 58 percent of the urban population. The remainder is dispersed in small and medium-sized towns ranging in population from around 5,000 to 50,000. Baraut, with a population of about 50,000, is an important market town and also the block headquarters. Most of the other twenty-three towns have populations below 20,000 (Uttar Pradesh, State Planning Institute, *Sankhyiki Patrika, Meerut*).

Agricultural Performance

A brief review of the background to the Green Revolution in Meerut reveals that the district presented a favorable environment for rapid modernization. The fertile soil and high water table made it possible for farmers to adopt the package of practices recommended for the HYVs. Consolidation of holdings in Meerut started in the early 1950s and by the mid-1960s was more than 80 percent completed; however, population pressure led to continuous subdivision of landholdings, so that the average landholding fell from 2.7 hectares in 1960–61 to 1.2 hectares in 1985–86 (Singh 1981; Uttar Pradesh,

Board of Revenue, *Agricultural Census of Uttar Pradesh).* A second round of consolidation was under way by the late 1970s.

Just as the district has enjoyed a long tradition of canal irrigation, its land and agriculture have remained largely in the hands of small owner-proprietors. The 1947 figures of the *Zamindari* Abolition Report showed that "in Meerut District large estates were few but there were an enormous number of petty proprietors . . . as well as the largest number of yeoman farmers in the whole of U.P." (Stokes 1978:218). With its favorable physical and institutional environment it is not surprising that Meerut adopted HYV technology with such alacrity.

In the mid-1960s the net irrigated area in Meerut was about 58 percent of the net cultivated area. Of the irrigated area, 54 percent was accounted for by canals, 21 percent by tubewells, and the remainder by traditional sources such as wells, ponds, and tanks. By the mid-1970s irrigation coverage increased to more than 86 percent. But more notable was the change in the nature of irrigation: more than 53 percent of the net irrigated area was served by tubewells. By the late 1980s irrigation coverage had increased further to 97 percent, with more than 70 percent accounted for by tubewells. The cropping intensity was then of the order of 214 percent (Uttar Pradesh, Board of Revenue, *Crop and Season Reports of Uttar Pradesh*; Uttar Pradesh, State Planning Institute, *Sankhyiki Patrika, Meerut).*

Wheat yields, which stood at around 1 ton per hectare on the eve of the Green Revolution, almost doubled within the next decade and had trebled by the mid-1980s. Cropping patterns shifted in favor of wheat, which replaced *rabi* pulses and oilseeds (mainly gram and rapeseed).

The agriculture of Meerut District is dominated by wheat and sugarcane, which together account for 64 percent of the total cultivated area. Fodder also plays an important role in cropping patterns, accounting for more than 20 percent of the total cropped area. Small amounts of coarse cereals, pulses, and oilseeds, together with potatoes, vegetables, and fruits, account for the remainder. Since the early 1980s the development of short-duration varieties of pulses and oilseeds has resulted in the reversal of the earlier decline.

The agricultural calendar includes three major seasons: *kharif* (the rainy season), *rabi* (the winter cropping season), and *zaid* (the dry summer season that follows *rabi).* Virtually all the land is double cropped, and many farmers cultivate three crops a year. These are usually cultivators with private means of assured irrigation who can back up their electric tubewells with diesel pumpsets in the summer season when interruptions in the power supply are frequent. Multiple cropping makes major demands on the availability of irrigation and requires timely agricultural operations. An increasing trend in the use of threshers for wheat and tractors for field preparation is apparent in Meerut. Three-crop rotations, such as wheat/maize fodder/sorghum fodder, or wheat/short duration pulses/sorghum fodder, are quite common; four-crop rotations, such as late wheat/green manure/fodder/early potatoes, are gradually gaining ground.

A farmer with 4 hectares or more commonly grows four crops a year on part of his land and plants the remainder in sugarcane. HYV wheat is harvested about April 20; two or three days afterward a leguminous green manure crop is sown; by June 25 it is plowed under and *cheri* (sorghum fodder) is planted, which is harvested in two months; by September 15 a crop of potatoes is sown which is harvested by late November, and the field is sown to wheat again.

Reasons for Selecting Meerut

The Green Revolution took early root in Meerut, and in many respects it had as strong an impact in Meerut as in the Punjab, the latter being synonymous with the Green Revolution in India. Table 4.1 compares some indicators of agrarian structure and agricultural performance for Meerut, western U.P., and Punjab. It also presents these indicators for the three case-study villages that are discussed in subsequent chapters.

The percentage of workers in agriculture is almost equal for Meerut and Punjab. However, the proportion of small and marginal landholdings in Meerut is almost double that in Punjab, where most landholdings are larger than 2 hectares. Population density in Meerut, at 708 persons per square kilometer, is almost twice that of Punjab. It is the difference in population pressure on land that accounts for the difference in the size of landholdings. Despite the disadvantage of small landholdings, Meerut has done well in overall agricultural performance. The irrigation coverage compares very favorably with Punjab, as do the wheat yields and the cropping intensities.[1]

There is not much information on the agricultural sector's linkage with the rest of the economy, and particularly on rapid agricultural

---

1. The cropping intensities are not strictly comparable because they have been estimated by counting sugarcane as a double crop. While this crop accounts for 31 percent of total cultivated area in Meerut, it constitutes an insignificant 1 percent in Punjab.

Table 4.1. Selected agricultural indicators for Punjab and western Uttar Pradesh, mid-1980s

Indicator	Punjab	Western U.P.	Meerut District	Rampur village	Izarpur village	Jamalpur village
Workers in agriculture as percentage of total work force, 1981	54	69	56	57	42	77
Small and marginal landholdings as percentage of total landholdings, 1980–81	39	82	78	62	88	83
Average area per landholding (ha), 1980–81	3.79	1.22	1.33	1.75	0.61	1.83
Net irrigated area as percentage of net crop area, 1985–86	86	77	93	100	100	100
Cropping intensity, counting sugarcane as a double crop, percentage, 1985–86	170	169	214	218	202	207
Sugarcane area as percentage of gross cropped area, 1985–86	1	13	31	30	20	24
Wheat area as percentage of gross cropped area, 1985–86	45	35	33	29	32	24
Yield of sugarcane (tons/ha), 1985–86	64.7	50.9	47.8	48.8[a]	40.4[a]	44.0[a]
Yield of wheat (tons/ha), 1985–86	3.5	2.5	3.3	3.5[a]	3.2[a]	3.4[a]
Yield of rice (tons/ha), 1985–86	3.2	1.8	1.8	—	1.4[a]	—

Sources: India, Ministry of Agriculture, Directorate of Economics and Statistics, *Indian Agriculture in Brief* (Delhi, various years); Uttar Pradesh, Department of Agriculture, Directorate of Agricultural Statistics and Crop Insurance, *Uttar Pradesh Ke Krishi Ankre* [Agricultural statistics of Uttar Pradesh] (Lucknow, various years).
[a]Yields are for 1988–89.

growth's effect on nonagricultural employment; however, the growth of the rural off-farm economies in Punjab, western U.P., and Meerut exhibit similar trends. Small and medium-sized towns are evenly dispersed throughout the countryside, and off-farm activities tend to be related to agroprocessing and technology-linked services. Both suggest that off-farm diversification is largely stimulated by the secondary effects of the Green Revolution.

Some might argue that the off-farm economies of Punjab, Haryana, and Meerut are driven by their proximity to the capital city, New Delhi. If such were the case, we would expect to find similar patterns of rural diversification in regions surrounding cities bigger than Delhi —for instance, Calcutta and Bombay. There is no such evidence.

Mechanisms of Income Diffusion

The mechanisms of income diffusion as they operate at the village and household levels are demonstrated by case studies in Chapters 5, 6, and 7. The impact of each mechanism is examined in the context of a village in Meerut District where its operation appears particularly prominent. The following section provides a broad overview of the secondary effects of the Green Revolution as they have manifested themselves in the changing pattern of rural activity in Meerut District.

Off-Farm Diversification

Given agriculture's limited capacity to absorb labor (Sarma 1981, Shand 1983), rural off-farm activities assume increased importance as alternative or supplementary sources of rural employment and income. Agriculture's stimulation of production from the nonagricultural sector has received scant attention by scholars, and when it has been noted, its role has been underplayed. In the words of Albert Hirschman, "Agriculture stands convicted on the count of its lack of direct stimulus to the setting up of new activities through linkage effects—the superiority of manufacturing in this respect is crushing" (1958:109–110).

John Mellor (1986, Mellor and Desai 1986) countered this view by emphasizing the strength of the indirect influence of agricultural growth on nonagricultural employment. He argued that accelerated agricultural growth of the kind generated by the Green Revolution leads to favorable local multiplier effects, which are manifested in the rising demand for labor-intensive goods and services. Evidence suggests that the major share of incremental income of peasant farmers is expended on agricultural inputs, consumer goods, and various rural services which are produced labor-intensively within the rural areas.

The diversification of the rural economy in Meerut lends support to Mellor's view. Off-farm activities in small manufacturing, agroprocessing, transport, services, and trade are characterized by low capital but high labor intensity and a wide geographical distribution. Such diversification also counteracts some of the inequalities that resulted from the initial impact of the Green Revolution technology, because a large number of the beneficiaries are landless and near-landless households, a formerly bypassed sector of the rural population.

Extent of Off-Farm Activity in Meerut. Rural off-farm activity constitutes an important component of the rural economy of Meerut. According to the 1981 census about 27 percent of the rural labor force was dependent on nonagricultural occupations for their livelihood (India, Office of the Registrar General, *Census of India*, General Population Tables). This figure almost certainly underestimates the extent of off-farm employment in Meerut. One reason for this is that the definition of *rural* excluded small and medium-sized towns (population 5,000–50,000). These towns invariably have close ties with their rural surroundings and their economies are largely fueled by providing essential goods and services to the rural population. A reasonable assumption, based on our findings, is that at least half the workers in these growth centers can be considered an extension of the rural sector. If the definition of *rural* were broadened to include these towns, the share of rural off-farm workers in the total rural work force would increase to more than 35 percent (Uttar Pradesh, State Planning Institute, *Sankhyiki Patrika, Meerut*). A second reason for believing that the census figure is an underestimate is that it did not count workers who participate in rural off-farm activities as a secondary occupation or on a part-time or seasonal basis. This omission may have caused the census to understate off-farm employment by another 10 to 20 percent. Taking into account both factors, it is reasonable to assume that at least 50 percent of the rural work force is engaged in off-farm activities as either a main or a subsidiary occupation.

Nature of Off-Farm Activity in Meerut. Off-farm occupations may be classified into wage employment and self-employment. The former can be further grouped into the public and private sectors. The private sector, both organized and informal, provides the bulk of employment to the rural poor. Enterprises in this sector can be broadly classified into the following categories:

- *Agroprocessing units:* sugar factories, cane-crushing units, cold-storage facilities, flour mills, rice mills, oil-pressing units, pulse decortication units, bakeries, and so forth

- *Forest-based units:* manufacture of paper, stationery, matchsticks, furniture, and the like
- *Agricultural input industries:* units engaged in the production of agricultural implements, Dunlop carts, trolleys for tractors, threshers, pumpsets, tubewell pipes and accessories, tires for a range of road vehicles, fertilizer, seed, and spare parts for agricultural machinery
- *Consumer goods units:* textile factories, cloth-printing units, electrical and electronic goods, motor parts, sporting goods, and so forth
- *Service and distribution centers* for agricultural inputs and consumer items
- *Trade, transport, and construction*

In the self-employed category the enterprise is generally a small family unit requiring few capital inputs. These include the following:

- *Agricultural units within the village: kolhus* (small cane-crushing units for making *gur*), *atta-chakkis* (small grain mills), leather work, and processing of spices
- *Nonagricultural rural industry:* handloom and powerloom weaving, cloth printing, soap making, blacksmithy and carpentry, scissors-manufacturing services, small workshops and garages for servicing agricultural machinery, cycle repair, welding units, electrical goods repair and maintenance, tailors, barbers, tea stalls, and restaurants
- *Transportation:* operating rickshaws, horse carts, *tempo-taxis*, and buffalo carts (with Dunlop tires)
- *Small trade and business:* vending fresh agricultural and poultry produce, running a family *parchoon* (grocery) shop, selling garments on a handcart

In the mid-1980s Meerut District had about 5,400 officially registered rural industries and manufacturing units—largely handlooms, handicrafts, small engineering units, and agroprocessing units—which generated employment for about 45,000 persons (Uttar Pradesh, State Planning Institute, *Sankhyiki Patrika, Meerut*). It is likely that a much larger number of rural micro-enterprises operating in the informal sector went unrecorded.

Impact of Off-Farm Activity on Agricultural Wages. With the increased diversion of rural labor into off-farm activities it would be reasonable to expect that wage rates of agricultural labor would exhibit an increase. In fact, between 1965–66 and 1988–89 the real wage rates of agricultural labor rose by about 30 percent (India, Ministry of Agriculture, *Agricultural Wages in India*).

Unlike the Punjab, Meerut and western U.P. did not experience a large-scale seasonal influx of migrant workers from eastern U.P. and Bihar. In Punjab, agricultural wages have consistently remained one and a half to two times higher than in Meerut (India, Ministry of Agriculture, *Agricultural Wages in India*). The demand for hired labor is relatively less in Meerut, and consequently agricultural wages there did not register increases as high as those in Punjab. There appear to be several reasons for this. First, small and marginal holdings constitute 78 percent of farm units in Meerut, compared with 39 percent in Punjab. This indicates a higher pressure on land and greater availability of agricultural labor in Meerut. Second, unlike in Punjab, where the major crops are wheat and paddy, the cropping pattern in Meerut is dominated by sugarcane, a crop that occupies the land throughout the year and is less labor-intensive. Third, with increased diversion of male workers to off-farm activities, a large number of female workers are now involved in agricultural operations in Meerut. Women are usually paid lower wages than men. Fourth, many workers are employed in off-farm activities only on a part-time or seasonal basis. In fact, workers do not travel long distances to find off-farm employment; most find it in growth centers within easy commuting distance and are therefore available for work during peak agricultural operations. Many such workers, however, find time to work only on their family farms.

Importance of Growth Centers. There were twenty-five towns in Meerut District in 1981. Of these, Meerut City, with a population of about 500,000, accounted for 58 percent of the total urban population. The remaining 42 percent was spread over twenty-four small and medium-sized towns with populations ranging between 5,000 and 50,000 (Uttar Pradesh, State Planning Institute, *Sankhyiki Patrika, Meerut*). The wide dispersal of these growth centers indicates that their economy is largely geared toward meeting the requirements of the rural population that surrounds them (Fig. 3.1).

Growth centers in Meerut can be broadly classified into three categories. The first and oldest category consists of rural market towns that expanded as a result of increased volume of agricultural production and trade. Second is the class of factory township that grew up around an agroindustry such as a sugar factory or a textile mill. The third category of towns developed around the nucleus of government block development offices that were established in the 1950s. All growth centers are characterized by a network of infrastructural facilities for agricultural inputs, transport, communication, education, health, and law and order. Each growth center serves a cluster of villages, so that most essential services are within 15 to 20 kilometers of even the most interior villages. Figure 4.1 indicates the distribution of villages according to their distance from basic services.

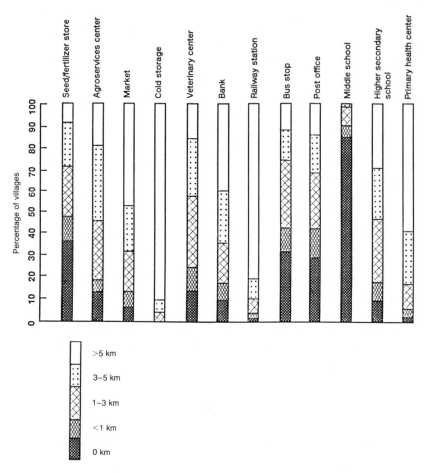

Source: Uttar Pradesh, State Planning Institute, Economics and Statistics Division, *Sankhyiki Patrika, Meerut* (Lucknow, 1989).

Figure 4.1. Percentage distribution of villages in Meerut District according to their distance from services, 1988–1989

Dairying as a Source of Income Diffusion

The demand for milk and dairy products has increased dramatically throughout India in recent years. Table 4.2 shows that the demand elasticity for these items is high among both rural and urban populations. This expansion has the potential for income diffusion because of dairying's labor-intensive nature and its ability to operate on a small scale with existing skills.

Any analysis of dairying is constrained by lack of data. Animal husbandry has always taken second place to crop production in India. The latter, which accounts for over 80 percent of the value of total agricultural output, has generally been used as the measure of agricultural performance. Hence the low priority accorded to livestock statistics.

Trends in Milk Production. Dairying has long been an integral part of the rural economy of Meerut. In 1920 it was observed of the districts of the Upper Ganges–Yamuna *doab* that "this was the tract where not only the province's best plow cattle but also 'the best class of milkers' were to be found" (Stone 1984:278).

In the early 1960s, the total daily milk production in Meerut District was a little over 300,000 liters, and the average yield per milk animal was about 1 liter per day (Uttar Pradesh, Department of District Gazetteers 1965). In the early 1980s the total production had

Table 4.2. Food expenditure patterns and income elasticities of demand in India, 1973–1974

Commodity	Percentage of total expenditure on food		Income elasticity	
	Urban	Rural	Urban	Rural
Cereals	40.2	59.5	0.21	0.48
Sugars and sweets	4.8	4.2	0.66	1.09
Pulses and products	4.9	5.0	0.53	0.83
Vegetables	6.6	5.0	0.70	0.70
Fruits and nuts	2.9	1.5	1.33	1.40
Meats, fish, and eggs	5.1	3.5	0.97	1.04
Milk and dairy products	13.6	9.6	1.06	1.41
Oils	7.7	5.0	0.70	0.85
Other foods	14.2	6.6	1.18	1.00
TOTAL	100	100	—	—

Source: Food and Agriculture Organization, *Income Elasticities of Demand for Agricultural Products* (Rome: FAO, 1983).

increased three and a half times to about 1.1 million liters, and the yield per animal had risen to about 3 liters per day (Agarwal 1988).

The threefold increase in yields was brought about partially through the spread of improved Murrah varieties of buffaloes and partly through improved feed. Although no data on animal feed are available, the increase in acreage under fodder crops suggests at least some improvement in animal nutrition. Between the mid-1960s and mid-1980s the acreage under fodder crops in Meerut rose from around 16 percent to 20 percent of the gross cropped area, an increase of about 20,000 hectares (Uttar Pradesh, Board of Revenue, *Crop and Season Reports of Uttar Pradesh*).

Involvement of Small Producers. Until recently several obstacles stood in the way of small producers widespread participation in dairying. Milk, a highly perishable commodity, could not be marketed as fluid milk from villages in the hinterland for lack of a well-developed infrastructure of roads and transportation. Milk cattle are expensive, and credit was not easily available. There is also an element of risk involved in the dairy enterprise.

The rapid improvement of the rural infrastructure, particularly the construction of rural link roads and the availability of cheap credit under government antipoverty programs, in the 1970s and 1980s removed some of these constraints and enabled larger numbers of landless and near-landless households to take up dairying. A study at the national level found that most milk produced in the country comes from small producers with one or two milk animals (Sarma 1981). Although no such statistics are available for Meerut, the composition of the cooperative milk societies that operate at the village level indicates that nearly 85 percent of the member-producers are landless, near-landless, or small landholders (Meerut, Chief Development Officer 1989).

Cooperative milk societies, which were introduced in the mid-1970s, provide an alternate marketing channel for village producers. The societies challenged the monopoly of the traditional *dudhiya*, or middleman, by offering higher prices to the producer-members, based on scientific testing of the fat content of milk. Although the cooperative marketing movement did not succeed in eliminating middlemen, who continue to purchase more than two-thirds of the surplus, it has resulted in a more competitive rural milk market. A further benefit to milk production came with the setting up of a cooperative milk-processing factory in Meerut in the late 1970s, with a capacity to process about 10 percent of the daily milk production in the district.

State intervention in the form of a special program designed to help landless and near-landless households to purchase milk cattle by providing loans and subsidies played a major role in the spread of dairying activity among the rural poor. The IRDP, which was started in 1980–81, had in eight years enabled more than 30,000 households in the less privileged category to purchase milk cattle (Meerut, District Rural Development Agency 1989).

Implications for Employment and Nutrition. It is difficult to assess the precise labor requirements of a household dairy enterprise. The activities comprise cutting, carrying and chaffing of fodder; feeding and milking; cleaning and maintaining the animals and their shed; marketing milk; and collecting dung and making dung cakes for fuel. Women and children are usually most involved in these activities. Thus dairying is a useful enterprise for exploiting surplus female family labor. In communities where employment outside the home is culturally taboo for women, dairying provides an excellent opportunity for gainful employment within the home.

Although most of the milk produced is marketed, one-fourth to one-third is generally retained for home consumption. Our conversations with small milk producers indicate that children in households with a milk animal have more access to milk than other children, implying that dairying may well lead to an improved nutritional status in children.

Cultivation of High-Value, Labor-intensive Crops

Of the three mechanisms of income diffusion, the cultivation of high-value crops—potatoes, fruits, and vegetables—is the least important in Meerut. About 6 percent of consumer expenditures for food in India is spent on vegetables. The income elasticity for vegetables is 0.7, while for fruit and nuts it is higher (Table 4.2). Since the bulk of vegetable production comes from small and marginal landholders, they are the group most likely to gain from the rising demand. Fruit, on the other hand, especially orchard fruit, is largely the domain of big landowners, and therefore enhanced demand is likely to benefit the larger category of farmers.

Emerging Trends in Acreage under Potatoes, Fruits, and Vegetables. The traditional crops of Meerut District are wheat, sugarcane, and fodder, and these three crops together account for more than 80 percent of the gross cropped area. Since the mid-1960s, however, the acreage under potatoes, fruit, and vegetable crops has shown a modest increase. Official statistics at the district level, based on village land records, indicate that potato acreage increased from around 3,000 hectares in the mid-1960s to about 12,000 hectares in the mid-1980s, yet it includes only 1.5 percent of gross cropped area (Uttar Pradesh, Board of Revenue, *Crop and Season Reports of Uttar Pradesh*). Although production of the crop is minor, discussions with farmers indicate that its value per unit of land ranges from one and a half to two times that of wheat, the other major *rabi* crop.

The acreage under fruit and vegetables also shows an increase; however, data on fruit and vegetables are open to question. In many official documents fruits and vegetables do not even merit a separate category and are grouped together with "other" crops, hence evidence of increasing acreage under vegetables is difficult to marshal. The two sources of data on vegetable crops provide differing accounts. The Land Revenue Agency estimated the fruit and vegetable acreage for Meerut at around 16,000 hectares in the mid-1980s (Uttar Pradesh, Department of Agriculture, *Uttar Pradesh Ke Krishi Ankre*), and the Horticulture Department put the figure at around 32,000 hectares (Meerut, Chief Development Officer 1989). Visual estimations during tours of Meerut District and discussions with cultivators suggest that the Land Revenue Agency records underestimate the figures and that reality lies closer to the Horticulture Department estimates. Again, although official statistics fail to disclose the fact, discussions with farmers across Meerut District indicate that fruit and vegetable acreage has increased significantly in the decade of the 1980s; both crops have gained in importance as alternate cash crops.

Implications for Employment. Compared with sugarcane, the traditional cash crop, potato and vegetable cultivation is more labor-intensive. Small and marginal landholders, who are able to cultivate vegetables on their small plots using family labor, are the major producers. The daily harvest and sale of the vegetable crop eases the household cash flow problem. Potatoes, on the other hand, although labor-intensive, are largely cultivated by medium and big farmers because the crop requires a heavy initial investment in the form of seed. Moreover, potatoes are a *rabi* crop and cannot compete with wheat on the fields of small and marginal landholders. However, some small farmers harvest the potato crop in December, earlier than its normal harvest period in February–March, and then sow a late crop of wheat. This reduces the yields of both crops but makes up in overall income by allowing two crops instead of one and fetching a higher price for the off-season potatoes. Irrespective of the size of the farm on which potatoes are cultivated, the labor-intensive nature of the crop generates employment for both family and hired labor. Labor use by crop in Uttar Pradesh, in man-days per hectare, is 195 for potatoes, 151 for sugarcane, 106 for maize, and 100 for wheat (Uttar Pradesh, Institute of Agricultural Sciences 1974). No such estimates exist for vegetable cultivation. However, it is a labor-intensive operation with most of the labor involved being female family labor.

Methodology and Data Collection

Data were collected during an eighteen-month period of field research running from March 1988 through August 1989. Regional and district-level data were obtained from census reports, published and unpublished statistics of the state government of Uttar Pradesh, and the U.P. Board of Revenue. This data base was supplemented by a large number of open-ended interviews with farmers, landless laborers, rural artisans, public representatives, social workers, and officials at the state, district, block, and village levels. These discussions provided a feeling for which groups were gaining from the second-generation effects of the Green Revolution and the broadening base of income generation.

Selection of Villages for Case Studies

Off-farm diversification and dairying are widely diffused activities, and households in almost all the 900 villages of the district are involved in these occupations to varying degrees. Hence the choice of villages to illustrate these two phenomena did not present a problem. Cultivation of high-value crops, especially potatoes, is not as universal, although vegetable crops are fairly widespread. The major potato-growing areas in the district are concentrated in Daurala, Jani Khurd, Meerut, Rajpura, and Kharkhauda blocks (Fig. 3.1). These five blocks together constitute about two-thirds of the total potato acreage. Almost all of the thirty-eight cold-storage facilities are located in these blocks.

Visual inspection of about twenty villages in the area and preliminary discussions with farmers and landless laborers as well as official agents helped to narrow the choice to one village for the vegetable and potato cultivation study: "Jamalpur"[2] in Kharkhauda Block. The village is about 5 kilometers in the interior and connected to the highway by a road that is part brick and part dirt. Once we had selected Jamalpur for our study of high-value crops, we chose villages for the off-farm diversification and dairying studies within the same general area for logistical reasons. "Rampur" village in Jani Khurd Block was selected to demonstrate the off-farm diversification strategy; more than 40 percent of its work force was engaged in nonagricultural occupations according to the 1981 census. "Izarpur" village, in Meerut Block, was chosen to represent dairying activities because a preliminary survey and discussion with farmers indicated that a large number of landless and near-landless households had dairying as their main or subsidiary occupation.

Another criterion for the selection of these villages was their size. Villages in Meerut District vary in population from 500 to 10,000, with an average of about 2,000. To keep the study manageable, small villages were selected. The populations of the three study villages range from 1,000 to 1,600.

Village Background Data. The only officially maintained data that enable us to observe changes over time at the village level deal with land use. The village land records are of two types: the *khatauni*, which documents land right, and the *khasra*, a record of the acreage of various crops cultivated year after year in the village, and hence an accurate depictor of changes in the cropping pattern; however, *khasras* are maintained for the previous twelve years only. For some villages the *khasras* of the original *bandobast*, or land revenue settlement, of the year 1951–52 are available. These were used, whenever possible, to observe changes in cropping patterns.

Our first step was to obtain background information about the study villages. To obtain a feel for the agrarian structure, production systems, demographics, and socioeconomic milieu of each village, data collected from the land records and agricultural censuses were supplemented by open-ended interviews with farmers, laborers, teachers, official agents, and political representatives in each village.

Preliminary Village Survey. Preliminary surveys of the study villages were conducted in April–May 1988 to obtain information on

2. Not the actual name.

landownership and households. These served as a basis for selecting the sample households interviewed in the subsequent surveys.

Household Sample Survey. Discussions with groups of farmers indicated that cultivators owning different-sized landholdings exhibit substantially different behavior with respect to income diffusion mechanisms. Size of landholdings was therefore made the basis for our subgrouping of rural households. Agricultural censuses describe changes in the distribution of landownership, tenancy, and irrigated area in terms of categories based on holding size, and we used subgroups, or categories, of households similar to those of the official agricultural census to facilitate comparison with the district-level data. We classified households on the basis of landownership as follows: landless (owning no land), marginal/near-landless (<1 ha), small (1–2 ha), medium (2–5 ha) and big (>5 ha).

The household population for each village from the preliminary survey was divided into these five subgroups. A stratified random sample of households was selected for estimating household incomes; thirty households were selected randomly in each of the three case-study villages. These sample households represented from 15 to 21 percent of the total household population in the villages.

A household questionnaire was administered twice to the randomly selected households, the first time to cover the *kharif* crops and the second for the *rabi* and *zaid* crops.

Data were collected on the following for the crop year 1988–89:

1. Name, age, and educational status of head of household
2. Family members, including their relationship to head, age, sex marital status, education, and nature of occupation
3. Details of land owned and tenancy
4. Crops cultivated in *kharif*, *rabi*, and *zaid*
5. Cost of inputs, including seed, fertilizer, irrigation, hired labor, hired machinery, or draft power
6. Yield and output of crops, marketed surplus, and marketing costs, if any
7. Farm assets, including livestock, tubewells and pumpsets, carts, and farm implements
8. Number of days worked as agricultural laborer and wages earned
9. Yield of milk cattle, production and sale of milk, and purchased feed

10. Details of income from off-farm employment and transfer income such as remittances, pensions, rents, and interest
11. Loan and subsidy obtained under government antipoverty programs and purpose
12. Ownership of consumer durables

The responses were used to estimate the income of the sample households. Estimated Lorenz curves and Gini coefficients were used to examine the degree of inequality in income distribution, and a decomposition of the Gini coefficient was performed to determine the effects of marginal changes in income by source to overall inequality.

Case-Study Household Sample Survey. In order to trace changes in the household economy of the landless and near-landless households and to identify factors and strategies that have enabled some households to achieve higher than average income levels, an additional five to fifteen households from these categories in each village were surveyed. These households were not selected at random. Rather, they were chosen from among the economically better-off households in their respective categories, and they especially illustrate the particular mechanism of income diffusion characteristic of that village. Each household was asked a variety of open-ended questions on past, present, and projected activities; response to government antipoverty programs; sources of credit; and farmer-community interactions. From their responses it was possible to build up detailed pictures of how the economies of these households have evolved.

Estimation of Household Income

Reliable income data are often difficult to acquire because of considerations of privacy, anticipated government interference, multiple income earners, and noncash earnings. Therefore considerable care was taken in the collection of information on income, and whenever possible the data provided by respondents were cross-checked by discussion with neighbors, the village headman, and the local-level government functionary.

Various types of income were identified, and, while the head of the household was usually the main respondent, information regarding dairying was cross-checked with female family members. Data were collected on crop income, agricultural labor income, animal husbandry income, off-farm income, and income from remittances, rents, interest, and pensions.

Crop Income. Crop income, or farm business income, was defined as income earned by family-owned factors of production: land, family labor, and capital. It was obtained by deducting paid-out costs from output value. Depreciation was not considered in calculating crop income.

Output valuation was made by accounting for total crop production on cultivated land during the *kharif, rabi,* and *zaid* seasons of the crop year 1988–89. Figures on actual weight of output of each crop were obtained directly from the cultivators. Since the questionnaires were administered during the harvest season it was possible to cross-check the actual weight of produce in about 20 percent of the sample households. Farmers' estimates regarding weight were found to be quite accurate. All by-products, such as wheat and rice straw, that have a market price were valued. By-products that do not command a market value were not valued.

Output actually sold was valued at the sale price given by the cultivator. A large proportion of the cereal crop output was saved for home consumption; this was valued at the average market price prevalent in the nearby market towns. For the *kharif* crops the average of market prices prevailing in November 1988 was used; for *rabi* crops we used the average for May 1989. Since all sugarcane was sold either to sugar factories, cane-crushing units, or the village *kolhu*, the actual sale prices for this crop were obtained in April–May 1989.

Valuation of input costs was made in accordance with the market prices whenever the inputs were obtained from the market. Fertilizer and chemicals purchased directly from the market or through the cooperative society were valued at prices actually paid by the cultivators. This was also true for seed purchased from the market. If seed was used from the previous year's production, however, an imputed value was assigned to it based on the market price prevailing at the time of sowing.

Irrigation was valued differently for those who purchased water and those who owned tubewells and pumpsets. In the case of canal water, the rate per unit cultivated area was fixed, irrigation rates for sugarcane being twice that for other crops. The rates for water purchased from private tubewells were per hour and varied according to the size of the tubewell. These rates were obtained from the respondent for each crop and cross-checked with the seller.

For those who had their own means of irrigation, the actual annual expenditure was estimated for the irrigation equipment—if operated on power, the actual electricity charges paid were used, if diesel operated, then cost of diesel fuel was used. In addition, the cost of equipment repair and maintenance was added to the total cost.

The labor input included only hired labor. As our definition of crop income included returns to family labor, no value was imputed to family labor. Wage rates included both cash and in-kind payments. Harvesting operations for cereals were paid for entirely in kind, the rate being about a one-twentieth share of the harvest. These were converted into cash equivalents using average market prices. Operations such as sowing and weeding were paid in cash plus a meal. An imputed value for meals was added to the cash value.

There was an active market in custom hiring of machinery and for draft power. The rental rate per unit of land for field preparation by tractors was usually fixed in the village, as was the rate for bullock rentals, although the latter have been almost replaced by tractors.

Other expenses included were land revenue, rent paid on land if any part of the land was leased, marketing cost of produce, and repayment of any loan installment on agricultural equipment.

Agricultural Labor Income. Agricultural labor income was evaluated on the basis of number of days worked in agricultural operations and the wages obtained for such work. The major problem in estimating agricultural labor income involved the accuracy of the number of days worked by various family members. The head of household as respondent was often unable to recall accurately the number of days worked by the various family members as paid agricultural labor. Efforts were made to cross-check information provided by the household head with the individuals, both male and female, who had actually done the work.

Animal Husbandry Income. The major source of animal husbandry income in the study villages was dairying. Dairying income was arrived at by estimating total milk production during the study period and deducting from it the cost of purchased inputs. Since the dairying was carried on almost entirely by family labor no hired labor costs were involved. The only purchased inputs were concentrates and fodder for the animals as well as any veterinary expenses. Fodder was valued only if purchased and if there existed a market for it. This was true for *berseem*, maize, and sorghum fodder for which the standing crop was sold and the purchaser could take several cuttings from the field. The valuation of the fodder crop was in terms of rupees per unit of land. Grasses, weeds, and other vegetation cut from the roadside and common land were not given cash values.

Data for average milk yields and number of days of lactation were collected to arrive at total milk production. Once again, this was based solely on recall and therefore subject to error. The information given by the head of household for dairying activity was cross-checked with the female household member most responsible for it. The actual milk sold was valued at its sale price. Milk for home consumption was valued at the same price at which the remainder was sold.

Off-Farm Income. Off-farm income was either from wage employment or self-employment. Under this category we recorded the nature and location of the activity and the net income accruing from the occupation. For regular wage employment the income was recorded in terms of a monthly salary or casual daily wages. If the latter, then the number of days of off-farm employment were recorded to obtain total annual income. For casual daily wage employment the actual number of days worked was a source of possible error because it was based on recall. On this point, information given by the head of household was cross-checked with the actual member involved in the off-farm activity. Income from self-employment was calculated after deducting paid-out costs.

Income from Remittances, Rents, Interest, and Pensions. Income from these four items was placed in one category, as they together constitute only between 5 and 14 percent of total income in the study villages. Cash remittances from family members received either in monthly or biannual installments were noted. Annual income from rental of land, buildings, and agricultural machinery was recorded under this category. This was based on information provided by respondents, and we were able to check in the village to ascertain whether the information was reasonable. Gifts in kind were valued at market prices. Although pensions were disclosed without any inhibitions, respondents were reluctant to disclose bank savings and interest. It is therefore likely that interest from savings and other financial investments has been underestimated in the case of the medium and big landholders. Wherever we felt that household income was being underestimated (by comparison with the general living standard of the household), some additional questions on household expenditures were asked of the respondent and he was asked to explain the difference. This usually elicited information not divulged earlier. This problem was encountered only with the wealthier households.

5

Off-Farm Diversification: Rampur Village

Rampur village highlights the off-farm diversification of the household economy as a mechanism of income diffusion. Our findings suggest that off-farm employment is a crucial factor in enhancing the economic well-being of the rural poor. The contribution of off-farm earnings to the landless and marginal households, besides improving their total income level, reduces the severe inequality of income distribution from crop cultivation alone. Access to off-farm employment has introduced changes in the rural labor market, which in turn are influencing cropping patterns.

The data also shed light on the nature of off-farm activities and indicate that the location of such activities is increasingly centered on rural market towns. These growth centers serve a cluster of villages and are located within easy commuting distance of each village, a factor that enables people to find off-farm employment while remaining domiciled in their village.

Profile of Rampur Village

As one drives west from Meerut City, vegetable fields give way to wheat and sugarcane interspersed with fruit orchards. *Buggis* (buffalo carts with tires) laden with sugarcane make their leisurely way to the cane crushers. Precariously loaded trucks thunder down the road on their way to sugar factories, with little concern for lesser traffic. The road is busy with buses, overloaded *tempo-taxis*, horse carts, tractor-trolleys, cyclists, and an occasional car or government jeep. High-tension power lines run across the flat landscape. The fields are dotted with tubewell houses. It is not uncommon to see a group of women harvesting a sugarcane field or carrying away *agolas* (bundles of sugarcane tops) to feed their animals. About 22 kilometers from Meerut City on the main road, nestled between mango groves and acres of sugarcane, lies the *abadi* of Rampur (Fig. 3.1). A pleasant village settlement with largely *pucca* houses and lanes paved with bricks, it has an air of well-being, confirmed by the appearance of the juvenile population, which looks healthy and well clad.

Bicycles and buffalo carts are the common mode of transport; about half a dozen motorcycles are also in evidence. More than a dozen television antennas proclaim the advent of the electronic age. On alternate Sunday mornings people congregate in neighbors' houses to watch every other episode of their favorite show, "Mahabharata." The situation is a telling comment on the state of the electricity supply, which is rationed due to the pressure of demand and is supplied to the rural areas in the state eight to ten hours a day—one week during the day, the following week at night.

The image of the traditional village women balancing pots of water on their heads is a thing of the past. Hand pumps in every courtyard provide drinking water. For those who cannot afford their own, there are four community hand pumps. Two *atta-chakkis* (small power-driven grain mills) have rendered obsolete the millstone and further reduced the drudgery of women.

A primary cooperative society and a cane cooperative society provide credit and marketing facilities to farmers in the village. The milk cooperative society runs a collection center in Rampur to enable its members to get remunerative prices. The center has helped in making milk prices more competitive, although more than 50 percent of landless and marginal producers continue to sell to the *dudhiyas*, or private milk vendors, in the village.

Rampur is connected by paved road to several market towns. Jani, the block headquarters town, is about 7 kilometers away. Dhaulri, Aminagar Sarai, and Siwal Khas, all bustling market towns, are within a radius of 20 kilometers from Rampur (Fig. 3.1). During the 1970s and 1980s these centers registered impressive growth in the trade, commerce, services, small manufacturing, transport, and construction sectors. Grain merchants, dealers in livestock, fertilizer shops, agroservice centers, electricians, mechanics, furniture makers, radio shops, photographers, tea stalls, barbers, and tailors have increased in number.

The 1981 census recorded a population of 1,044 persons and 164 households in Rampur. Our preliminary village survey conducted in 1988–89 found 190 households. There was no evidence of any significant emigration from the village. The only instance was in households with sons who are serving in the armed forces or the police and, who eventually intend to return to the village. Jats account for more than 60 percent of all households in Rampur and own more than 85 percent of the cultivated land. The Jats are widely recognized as excellent farmers and have played a major role in ushering in the Green Revolution in western U.P. The procedure for selection of sample households was discussed in Chapter 4. The breakdown of the sample for Rampur is presented in Table 5.1.

Changes in Agrarian Structure

The average operational holding in Rampur was 2.27 hectares in 1970–71; it fell to 1.75 hectares a decade later, primarily as a result of population pressure and subdivision of landholdings. The 1988–89 sample survey estimates that land owned per household is now 1.37 hectares. The village has a highly skewed distribution of landownership; 43 percent of households own no land at all, although there was an improvement between the last two censuses (Fig. 5.1). The preliminary village survey of 1988–89 indicates that only one household owns more than 5 hectares, and this is because there is joint cultivation by an extended family. The class of big landholders has been

Table 5.1. Total households and composition of Rampur sample, 1988–1989

Household	Village total		Sample size
	Number	Percentage	
Landless	82	43	13
Landowning	109	57	17
Marginal/near-landless (<1 ha)	44	23	7
Small (1–2 ha)	45	23	7
Medium (2–5 ha)	19	9	3
Big (>5 ha)	1	1	—
TOTAL	191	100	30

Source: Rampur preliminary village survey, 1988–89.

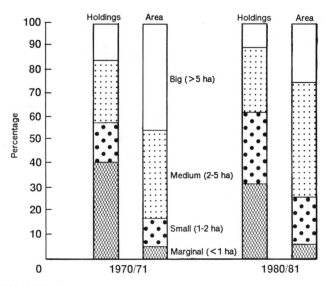

Source: Table A5.1.

Figure 5.1. Percentage distribution of landholdings and land owned in Rampur, by landholding group, 1970–1973 and 1980–1981

reduced to medium and small landholders. If the pressure on the land continues, it is likely that a decade or so may witness the extinction of the medium landholder class also.

Changes in Agricultural Performance

As Figure 5.2 shows, wheat, sugarcane, and fodder are the major crops in the village, accounting for almost 80 percent of the gross cultivated area. In the pre–Green Revolution period pulses and oilseeds occupied an important share of the cropped area, but they have been replaced by HYV wheat and sugarcane. In the 1980s, vegetables and fruit, especially mango orchards, began emerging as cash crops in the village. The increasing acreage under mango orchards—a labor-saving crop—reflects the labor shortage in the village that has resulted from diversion into off-farm activities.

Changes in Employment Patterns

The composition of the work force in Rampur suggests a high degree of participation in nonagricultural activities (Fig. 5.3). Indeed,

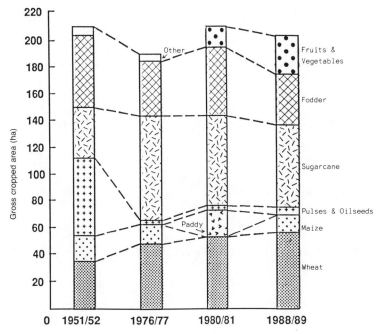

Source: Table A5.2.

Figure 5.2. Gross cropped area under major crops in Rampur, 1951–1952, 1976–1977, 1980–1981, and 1988–1989

the number of workers in nonagricultural activities registered an increase of more than 60 percent in the 1970s, while workers in agriculture rose by only 30 percent. It is expected that the 1991 census will show an even higher percentage of workers engaged in nonagricultural employment.

Of the households sampled in 1988–89, the landless were the main participants in noncrop activities; about 70 percent of these households had off-farm activities as their main occupation. For marginal households this figure was more than 40 percent. Small and medium farmers reported farming as their major source of income.

Increasing Importance of Off-Farm Activities

Both "push" and "pull" factors have contributed to the growth of off-farm employment. The major "push" factor has been the inability of marginal landholdings to yield an adequate living from agriculture,

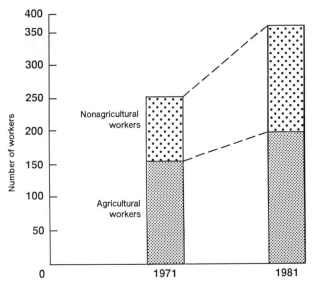

Source: Table A5.3.
Note: Nonagricultural includes animal husbandry, household industry, manufacturing, construction, trade, transport, and services; agricultural includes cultivators and agricultural laborers.

Figure 5.3. Composition of Rampur work force, 1971 and 1981

despite the major breakthrough in agricultural technology. The demand for agricultural labor, after rising in the early years of the new wheat varieties, has now begun to fall because of the growing number of small and marginal landholdings, where hired labor may not be needed except for harvesting operations. In Rampur the predominance of sugarcane and fodder in the cropping patterns, which together accounted for nearly 65 percent of total cultivated area in the mid-1970s, further dampened labor absorption in agriculture, so that landless and near-landless households are now unable to subsist on crop cultivation and agricultural labor alone.

The "pull" is exerted as a result of the second-generation effects of the Green Revolution. The agricultural growth of the mid-1960s increased rural incomes and generated consumer demand for nonagricultural goods and services, thereby stimulating linkages with rural-based activities of manufacturing, distribution, and services. Technology-linked occupations such as servicing of tubewells and tractors, custom hiring of machinery, construction, and rural agroindustries are very much in evidence. It is through this diversification that the benefits of the Green Revolution technology are diffusing to the marginal and landless class. While there are no quantitative benchmarks for comparison, the interviewees indicate that the years since the mid-1970s have witnessed an exceptional increase in off-farm activities.

Appeal of Off-Farm Employment

Landless and near-landless households find off-farm activities especially attractive because they provide an opportunity to utilize family labor resources. Another factor in favor of off-farm employment is the higher wages of these occupations compared with agricultural labor. A third reason has to do with the fact that these activities can be performed within easy commuting distance of the village. Most activities can be carried on in the village itself or in nearby growth centers.

Nature of Off-Farm Activities in Rampur. Off-farm activities involve both self-employment and wage employment. In Rampur, the sample survey suggests that rather more (56 percent) of the jobs fall into the latter category and rather less (44 percent) into the former. More than 70 percent of nonfarming occupations are carried out within a radius of 20 kilometers from Rampur. For the remainder, the workers had to commute to Meerut City or to the neighboring district of Ghaziabad, a distance of about 22 to 25 kilometers.

Table 5.2 indicates the types of activities engaged in. The self-employed jobs range from unskilled to skilled activities in trade (*parchoon* shop, milk vendor), transport (buffalo cart, *tempo-taxi*), services (mechanic, electrician, washerman, photographer), and rural industry. Wage employment similarly covers a wide range of activities and can be subclassified according to whether the employment is with the private or the government sector. In Rampur the private sector predominates. Jobs requiring minimal skills and low to medium capital investment are common in landless households. For medium and big farmers, income from hiring out their farm assets constitutes the most important supplementary source of income. Whereas landless and marginal households maximize returns to family labor, big farmers optimize their use of capital assets.

Table 5.3 presents the breakdown of off-farm income by various occupations. Off-farm income for the total sample is almost equally distributed between agroindustry, transport, and services, with trade accounting for around 16 percent of the total. The landless class ben-

Table 5.2. Off-farm income sources in Rampur, 1988–1989

Self-employment	Percentage	Wage employment	Percentage
Trade	8	Trade	8
Transport	8	Transport	16
Services	24	Construction	12
Rural industry	4	Private factory	16
TOTAL	44	Government	4
		TOTAL	56

Landholding group	Self-employment	Percentage	Wage employment	Percentage
Landless	Parchoon shop	4	Grain merchant	4
	Milk vendor	4	Fruit contractor	4
	Buffalo cart	4	Bus conductor	4
	Mechanic	4	Bus driver	4
	Electrician	4	Construction labor	12
	Washerman	4	Private factory	4
			Government (telephone linesman)	4
Marginal (<1 ha)	Rural industry	4	Tractor driver	4
	Mechanic	8	Bus driver	4
	Tempo-taxi	4	Private factory	8
Small (1–2 ha)	Photography shop	4	Private factory	4
TOTAL		44	TOTAL	56

Source: Rampur household sample survey, 1988–89.

efits more from employment in trade and transport, while for the marginal category more than one-third of income accrues from each of the agroindustry and transport sectors.

Higher Wage Rates. The sample survey indicates that the lowest paid casual jobs are in construction, where the average daily wage rate is Rs 30. The average for agricultural labor is Rs 20. For semi-skilled and skilled labor, off-farm employment generates between Rs 35 and Rs 60 per day.

Impact of Off-Farm Employment on Crop Cultivation

Prima facie it might appear that off-farm activities attract labor from farming and reduce both the quantity and quality of inputs in agriculture, especially since the more remunerative off-farm opportunities are those requiring better skills and education. However, the findings from Rampur suggest that no conflict exists between off-

Table 5.3. Composition of off-farm income in Rampur, 1988–1989

Income source	Landless		Marginal (<1 ha)		Small (1–2 ha)		Total	
	Rupees	Percentage	Rupees	Percentage	Rupees	Percentage	Rupees	Percentage
Agroindustry	9,000	7.8	28,001	36.7	5,999	100	43,000	21.7
Trade	32,600	28.1	—	—	—	—	32,600	16.4
Transport	25,540	22.0	26,580	34.8	—	—	52,120	26.3
Services	20,672	17.8	21,740	28.5	—	—	42,412	21.4
Construction	16,200	14.0	—	—	—	—	16,200	8.2
Government	12,000	10.3	—	—	—	—	12,000	6.1
TOTAL	116,012	100	76,321	100	5,999	100	198,332	100

Source: Rampur household sample survey, 1988–89.

farm and agricultural activities, indicating that slack exists in the labor market and that surplus, underutilized family labor can be drawn out of agriculture with little or no decline in output. A noticeable trend is the increase in women working in agriculture. The 1971 census recorded no female workers in Rampur; in 1981 they constituted 13 percent of all workers. Our 1988–89 survey indicates that the percentage of female workers has risen to more than 35 percent of the work force.

The following data from the 1988–89 household sample survey suggest that the diversion of labor into off-farm activities has not adversely affected the farm productivity of landholders in the marginal category:

Size group	Cropping intensity (percentage)	Net income per hectare (rupees)
Marginal	251	11,508
Small	180	10,083
Medium	232	10,032

Cropping patterns by farm size provide an indication of how the different landholding groups have responded to the diversion of workers into off-farm activities (Fig. 5.4). That sugarcane is the major cash crop for marginal and small farmers as well as for those in the medium category is an indication of the extent to which these households are trying to save labor. Trends toward potatoes and vegetable crops, which are labor-intensive, are negligible in Rampur. On the other hand, conversion of agricultural land into mango orchards by small and medium farmers during the 1980s, not only in Rampur but in several villages in the vicinity, suggests an adjustment of cropping patterns in favor of labor-saving crops.

Comparison of agricultural wages over time at the village level is not possible. However, district-level data on agricultural wages suggest an increase of about 30 percent in the real wage rate between 1965–66 and 1988–89. In the same period the wage rate of a skilled worker (blacksmith) rose more than 100 percent.

Factors Influencing Off-Farm Employment

The types of off-farm activities pursued in Rampur have been influenced by the infrastructure of the village and its proximity to nearby growth centers, and also by their input requirements.

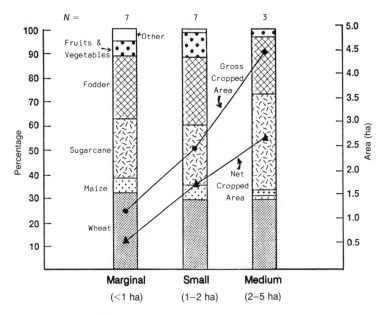

Sources: Tables A5.4 and A5.5.

Figure 5.4. Cropping patterns in Rampur, by landholding group, 1988–1989

Infrastructure Development. The infrastructure that developed during the 1970s and 1980s has played a vital role in the expansion of rural off-farm activities. Rural electrification has been the key factor in modernizing traditional rural-based processing and manufacturing activities. Rural power supplies, although erratic, have made possible the operation of processing industries, powerlooms, lathes, and farm equipment, which in turn has generated a whole new service sector of mechanics, electricians, and welders. Roads and communications facilitate commuting for employment and the movement of goods and services to and from market towns.

Growth Centers. About 70 percent of all off-farm activities are located either in Rampur village or within a 20-kilometer commute. The remaining 30 percent require further travel, with buses, trucks, and *tempo-taxis* being the common modes of transport. A good network of roads and government and private transport makes this possible. When asked why they do not migrate to bigger towns and cities, our respondents indicated that they feel their present net incomes are not much lower than potential incomes from jobs in big cities, and that in the village it is possible to obtain many basic needs at a cost much below that in town. Besides, community and kinship ties can still be counted upon as insurance against unemployment and under-employment, although these ties are gradually weakening. Family support systems provide social security in times of natural disasters, medical or other emergency, and especially in the case of the loss of a household head. Their decision to remain domiciled in the village is based on the calculated opportunity cost involved in living in the city.

Growth centers are increasingly becoming the focus of economic activity, especially in the service sector, for the villages clustered around them. Jani, Dhaulri, and Aminagar Sarai are all within 20 kilometers of Rampur. Dhaulri is a market town fast developing into a major fruit-marketing center. The block development office and the primary health center are located in Jani. Aminagar Sarai combines a market town with small manufacturing units as well as government facilities; a secondary school, post and telegraph office, veterinary services, seed store, bank, and fair price shop are to be found there. All these growth centers have attracted considerable private enterprise, especially in trade, services, and small manufacturing.

Availability of Skills and Capital. The level of earnings in off-farm activities is determined primarily by two factors: *capital* and *skills*. Table 5.4 presents a breakdown of off-farm activities by skill and capital requirements, as found by the 1988–89 household sample survey and observed in Rampur and adjacent villages.

CATEGORY I. Category I enterprises require low levels of skill and capital and yield low returns. The average daily net return from activities in this category ranges from Rs 10 to Rs 20 per day. The average daily wage for agricultural labor in Rampur is Rs 20.

While earnings from this category of enterprises are generally low, some households engage in them to take advantage of underemployed members whose opportunity cost is zero. Generally children, women, and older people take up such activities. When household heads engage in services such as washing clothes and sweeping, or female members in domestic work, the payment that they receive in kind is usually much below the market wage. However, these activities are valued more in terms of their ability to gain access to factors of production through interhousehold linkages. Individuals performing services for big and medium farmers are given land on lease for cultivation, priority in employment at harvest time, and access to the landowner's fields for gathering fodder and fuel.

Table 5.4. Skill and capital requirements of off-farm self-employment opportunities in Rampur and vicinity, 1988–1989

Capital	Skills and education	
	Low	Medium to high

Low	*Category I*
	Petty roadside trade
	  Vegetable/fruit vending
	  Tea shop
	Services
	  Washermen
	  Sweepers
	  Potters
		Category II
		Services
		  Barber
		  Blacksmith-carpenter
		  Chair caners
		  Basket makers
		  Lock makers
		  Tailor
		  Mason

Medium
to
high
	Category III
	Trade
	  Parchoon shop
	  Milk vending
	Transport
	  Buffalo/horse chart
	Custom hiring machinery
	  Tubewells/pumpsets
	  Threshers
	  Tractors
	Moneylending
	Remittances
		Category IV
		Trade
		  Grain trading
		  Milk/dairy products vending
		  Shops selling agricultural
		    inputs in market towns
		Transport
		  Tractor-trolleys
		  Tempos, trucks
		Services
		  Electrician
		  Mechanic
		  Welders
		  Radio/TV repair
		  Photographer
		Agroindustry
		  Grain mills
		  Oil pressers
		  Sugarcane crushers
		  Brick kilns
		  Sewing thread

Source: Rampur household sample survey and field notes, 1988–89.

CATEGORY II. These activities require low capital but medium to high skills. The daily net return from activities in this category ranges from Rs 30 to Rs 60. These are traditional occupations that existed in the feudal system under *jajmani*, the patron-client relationship. With modest capital investment in improved tools and equipment, these occupations have been adapted to serve a changing demand. Increased clientele and a larger volume of business are making these services economically viable. For example, caning chairs as an occupation in villages and rural areas was unheard in the mid-1970s. People sat on string cots or on mats on the floor. Today, there is enough demand in the rural areas for caned chairs. This enables a family of three to four members with the requisite skill to supplement their household income significantly.

CATEGORY III. Activities in this category are characterized by low skills but medium to high capital. Daily earnings under the trade and transport subsections range between Rs 30 and Rs 60. It is difficult to estimate earnings on a daily basis for activities such as custom hiring and moneylending. Under trade and transport the sample includes several households that had set up small shops with either private capital or cheap credit under the government loan and subsidy program. These enterprises usually require investments ranging from Rs 5,000 to Rs 10,000. No major skills are necessary.

CATEGORY IV. These enterprises are marked by a higher level of both skills and capital. The lowest daily net return in this category is about Rs 50. Investment in these types of enterprises can range from Rs 20,000 for a grain mill to Rs 45,000 for a *tempo-taxi* to more than Rs 1 million for a brick kiln.

Most landless and near-landless households operate in Categories I, II, and III. Very few are able to set up enterprises in Category IV. The primary constraint is capital. For the landless and marginal households, whose annual incomes average about Rs 16,000 and Rs 25,000, respectively, it is difficult to accumulate sufficient savings to finance a Category IV enterprise. Nevertheless, several of our case-study households demonstrate that it is possible to make the transition into Category IV, partly through household savings, partly through private credit, and partly through entrepreneurial skill.

Income Diffusion in Rampur

There has been considerable diversification of the economy in Rampur. Crop cultivation constitutes less than 40 percent of the village income. Off-farm earnings account for more than 30 percent, followed by dairying. The rapidly increasing participation of landless and near-landless households in off-farm activities has resulted in a marked improvement in both the absolute and relative incomes of the poorer households.

Diversification of Household Income

Figure 5.5 presents the household income estimates for Rampur village by size of landholding group for the crop year 1988–89. It is

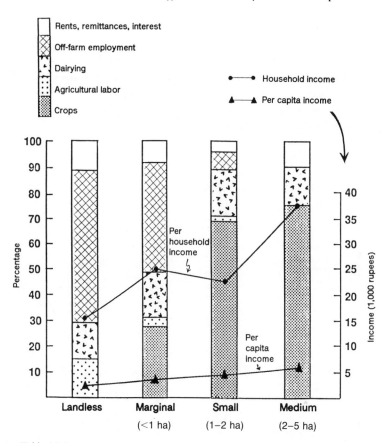

Source: Table A5.6.

Figure 5.5. Amount and sources of income in Rampur, by landholding group, 1988–1989

clear that off-farm earnings play a major role in the income of landless and marginal households. In the case of the former, off-farm activities account for nearly 60 percent of total income, whereas for the latter their contribution is more than 40 percent. The most striking feature of off-farm income is its tendency to equalize average household income across size groups. The figure shows that while the share of crop income varies directly with farm size, off-farm income suggests an inverse relationship. The overall impact of off-farm income has been to compensate in part for the large variation in crop income between groups. While the crop income of medium households is almost twelve times higher than that of the landless households, the total income is only two and a half times higher.

Comparison with the Poverty Line. To appreciate the importance of off-farm income to landless and marginal households, it is appropriate to recall that the official poverty line at the time of our survey was around Rs 8,600 per household of five members, or Rs 1,700 per capita. If landless households had not engaged in off-farm employment, their income would have been below the poverty line. The situation of marginal farmers would also have been precarious.

Implications for Income Distribution

Figure 5.6 presents the Lorenz curves and the corresponding Gini coefficients for crop income, noncrop income, and total income.

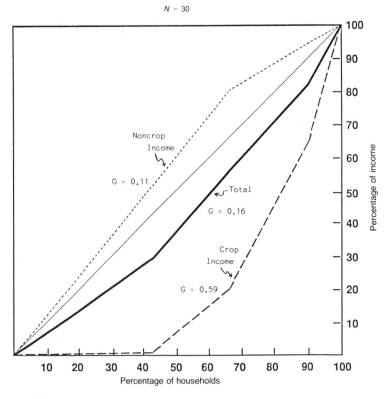

Source: Table A5.7.

Figure 5.6. Lorenz curves for Rampur, by landholding group, 1988–1989

Given that 43 percent of the households are landless, the inequality in income from crops is not surprising.

Table 5.5 presents the results of the decomposition of the Gini coefficient for total income for Rampur village. Column 2 presents the share of each income source in the total income of the sample households. The Gini coefficient in column 3 indicates the distribution of each income source. Column 4 presents the Gini correlations between each income category and total income. Agricultural labor and off-farm employment have a negative correlation with total income. This indicates that the landless and near-landless groups, at the lower end of the income distribution, receive a greater share of their income from these sources.

The importance of the Gini correlation is evident when we compare the contribution of each income source to village income inequality (column 6). Crop income accounts for the largest share of inequality, and agricultural labor and off-farm employment tend to counteract inequality caused by crop income.

To estimate the impact of small percentage changes in income from a specific source on total inequality we examine the elasticity of inequality in column 7. A 1 percent increase in crop income would exacerbate inequality by 0.71 percent, while a similar increase in off-farm income would reduce income inequality by 0.52 percent. It is clear that off-farm earnings, besides improving the income levels of the rural poor, play a major role in reducing income disparities between landholding groups.

Table 5.5. Decomposition of income inequality in Rampur, by income source, 1988–1989

Income source	Income share (S)	Gini of source (G)	Correlation with rank of total rank (R)	Share of inequality		Elasticity of total inequality by income source
				SGR	Percentage	
(1)	(2)	(3)	(4)	(5)	(6)	(7)
Crop income	0.38	0.59	0.80	0.18	109.0	0.71
Agricultural labor	0.06	0.49	−0.79	−0.02	−13.7	−0.19
Dairying	0.17	0.20	0.93	0.03	20.1	0.03
Off-farm employment	0.31	0.38	−0.29	−0.03	−20.9	−0.52
Rents, remittances, pensions, interest	0.08	0.20	0.54	0.01	5.4	−0.03
TOTAL	1.00	0.16	1.00	0.16	100	—

Source: Rampur household sample survey, 1988–89.

Household Economic Strategies

The landless and marginal households in Rampur constitute 67 percent of all households, and those in the small and medium categories account for the remainder. Since off-farm activities contribute a major share to the incomes of the landless and marginal households, while crop cultivation is the most important source for the small and medium farmers, it is clear that the former follow economic strategies significantly different from those of the latter.

Resources and Objectives

Landless and Marginal Households. Landless and marginal households pursue multiple objectives, which include food security, risk minimization, maintenance of cash flow, and maximization of family welfare. Welfare is a complex goal comprising a mix of monetary and nonmonetary, quantifiable and nonquantifiable factors. Perceptions of welfare affect the goals of households and in turn influence the allocation of resources. Social and cultural factors play a vital role in household decision making. Housing, for instance, is conventionally considered a form of consumption rather than a means of generating income. In the village, however, construction of a house or conversion of a mud house into a brick one is often the primary objective of a household. In fact, many a loan granted under the government program for acquiring productive assets is used for house construction. There exists a social prestige in the ownership of a *pucca* house, not measurable in economic terms, that takes precedence over the economic gains accruing from the purchase of a milk animal.

Labor is the most abundant resource among landless and marginal households. Their major objective, therefore, is to employ it in various uses as efficiently as possible. This implies that households seek to maximize the subjective return to the labor of members, so long as they are simultaneously able to achieve food security, maintain a flow of cash, and minimize risk.

The multiple objectives of the landless and marginal households are manifest in the diversified production strategies they adopt. Government intervention or technical change is often concerned with influencing one aspect of the several strands that make up a household's strategy, whereas the households themselves are concerned with maintaining or enhancing their entire welfare. This might result in their rejecting what might otherwise seem to be advantageous. For example, the establishment of a milk cooperative collection center in Rampur did not evoke the expected response from all marginal and landless households, although it offers higher prices for milk. Some producers continue to sell to local private vendors. Although these middlemen purchase milk at rates lower than the cooperative milk center pays, they provide credit for consumption purposes and purchasing fodder. Thus they help the households achieve their objectives of supplementing income, maintaining a cash flow, and minimizing risk, but compromise on the single objective of profit maximization from the livestock (capital) resource.

Off-farm diversification as a household strategy fits in with the multiple objectives of the rural poor. It provides an additional source of income and helps reduce risk. In times of drought, when marginal farmers and agricultural laborers have traditionally been the most adversely affected, households with nonagricultural earnings suffer less than those with no alternative sources of income.

Off-farm earnings ease the cash flow situation and provide the means to generate surpluses to be invested in both farm and off-farm asset creation. We observed that relatives and moneylenders were willing to provide credit more readily to households with some off-farm income source.

Off-farm diversification as a strategy facilitates a fuller utilization of household labor throughout the year. Sometimes the occupations pursued appear to generate minimal or no surpluses. Yet, from the household's point of view these activities are worthwhile because the marginal returns are greater than the marginal real cost of labor in periods when there is little or no alternative employment. Children or older household members vending fruit or corn-on-the-cob by the roadside are examples of such activities.

Small and Medium Households. The resource position for small and medium farmers is more flexible. Their land base and accumulated farm surpluses from the adoption of Green Revolution technology have enabled them to enhance their well-being within the crop sector. The objective of these households is to maximize returns to land, capital, and family labor after meeting their food requirements. This is reflected in their cropping patterns. Cultivation has become increasingly market oriented and makes use of hired labor as well as machinery. None of these households hires out members for agricultural labor. They supplement crop income by the sale of irrigation water and the custom hiring of pumpsets, threshers, and tractors.

Gaining Access to Capital

Of the landless and marginal households surveyed in Rampur, 50 percent had availed themselves of a subsidized loan under the IRDP during the preceding ten years. However, the major source of credit for these households remains the private moneylender. In the past, *arthias*, or grain dealers in market towns, gave advances to farmers, who in return sold their produce to the *arthias* at below market prices. For the landless who had nothing to pledge, big landlords were the primary source of consumption loans. The borrower worked off the debt by providing labor for the landlord. A third alternative was the regular moneylender, who charged interest rates between 12 and 36 percent, with the farmer usually pledging his wife's jewelry.

Today, the *arthia* and the big landlord as moneylenders have virtually disappeared. In their place has emerged a new institution for private credit, the village milk vendor, who provides advances not only for dairying costs but also for consumption. With the increasing importance of dairying as a subsidiary occupation, more than 90 percent of households possess milk cattle, and the milk vendor is able to recover his loans with interest through purchase of milk at rates 15 to 30 percent lower than that offered by the cooperative milk collection center.

Ten percent of the households surveyed in Rampur reported borrowing from relatives to start an enterprise. Another 10 percent pawned or sold gold or silver ornaments. Accumulated savings from wage employment, remittances from migrant family members, and pensions and gratuities received by retired employees contributed significantly to capital formation in more than 80 percent of the households surveyed.

Private finance companies, which until recently catered only to the urban population, have spread their operations into rural areas. Their procedure is simpler than that of commercial banks, but their interest rate is almost 4 percent higher. Since they advance loans up to Rs 30,000 without undue emphasis on collateral, they do a brisk business in the transport sector. Of the several *tempo-taxis* in the Rampur area, most were financed by these private companies.

Under the government-sponsored IRDP, households below the poverty line are selected to receive loans at interest rates 3 percent lower than the market rate and with from 33 to 50 percent of the loan forgiven as a subsidy. The beneficiary under this program is expected to acquire some productive asset such as a milk animal, buffalo cart, or implement with which he can generate earnings to improve his household income level and repay the loan in installments over a period of three to five years. Several beneficiaries complained that the procedure for procuring an IRDP loan is lengthy and cumbersome and involves "leakages" on the order of one-third to half of the amount of the subsidy due to them. Some households were reluctant to apply after hearing of these problems from their neighbors. The selection procedure also came under criticism because some farmers felt that the poorest and most deserving among them have been ignored. The faults are not all on one side. There are cases of misutilization of loaned funds and a fairly high percentage of default on repayment. In the majority of cases in our sample, however, the asset created by the loan still existed and had helped in increasing household income.

Gaining Access to Skills and Education

A comparison between the level of formal education attained by the heads of households and their children among the sampled households indicates that the number of high school diploma holders has increased from 6 to 39 percent, and more than 70 percent among the younger generation have passed class 6. While there are no college graduates among household heads, 15 percent of the younger group hold B.A. degrees. Although education levels in the village have improved, there is a great deal of disillusionment with formal education; it raises aspirations for white-collar jobs but does not guarantee employment in such activities. Our case-study households include several high school diploma holders who could not procure a job and reverted to farming.

Vocational education has not received much attention so far. The major strategy of the more successful households has been to acquire professional skills and trades through the traditional method of apprenticeship. The government-run TRYSEM (Training Rural Youth for Self-Employment) program, which was started with the objective of imparting such skills to rural youth, has not had any significant impact in Rampur.

Case Studies

The case studies discussed next illustrate the strategies adopted to increase off-farm earnings. In addition to case studies from Rampur, some examples are taken from Sitapur, a village in the neighboring development block that exhibits the same trends in off-farm diversification as does Rampur.

Landless Households

We recorded the detailed experiences of five landless households. The methodology for their selection and the data collection process employed are treated in Chapter 4. These households were not selected at random; rather they were chosen because, given their circumstances, they have done rather better than the average.

Figure 5.7 presents the income profile of each of three case-study households of Rampur versus the average for the landless category in the village. Landless households constitute slightly less than half the households in Rampur. On average their annual income is on the order of Rs 15,000, and their per capita income is around Rs 2,300. Their entire income is derived from noncrop sources. Off-farm employment accounts for a bit less than 60 percent of the total, and

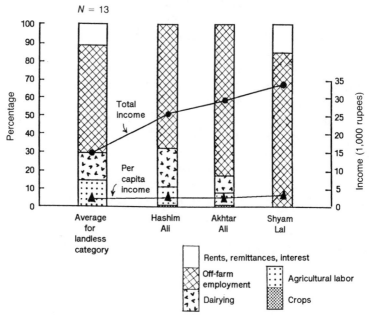

Source: Table A5.8.

Figure 5.7. Amount and sources of income for landless case-study households in Rampur, 1988–1989

agricultural labor, dairying, and rents and remittances each contribute a little more than 15 percent.

The three case-study households have a considerably higher than average household income. Per capita incomes, on the other hand, do not show as wide a variation from the average, indicating that household size is one of the factors determining household income. Indeed, the number of members in the Hashim Ali, Akhtar Ali, and Shyam Lal households are, respectively, 10, 9, and 10, contrasted with an average size for landless households of 6.6 members. The ability of family members to combine resources and cooperate within the extended family is influential in determining the level of household income.

Figure 5.8 presents the income profiles of the two households in Sitapur in the context of the average landless household in that village. Landless households account for almost two-thirds of the households in Sitapur. The annual income for the average landless household is around Rs 16,000, and per capita income is more than Rs 2,100. Off-farm activities account for about 65 percent of total income, dairying contributes one-fifth, and agricultural labor makes up another one-tenth; crop cultivation on a sharecropping basis plays an insignificant role.

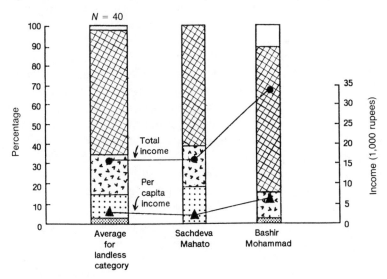

Source: Table A5.9.

Figure 5.8. Amount and sources of income for landless case-study households in Sitapur, 1988–1989

Of the two case studies, Sachdeva Mahato's household income is almost at par with the average, and the contribution of the various income sources also seems identical. However, Sachdeva expects the situation to improve as soon as his oldest son, who has just received a B.A. degree, gets a job.

Bashir Mohammad is one of the success stories of Sitapur. His household income is double that of the average and the per capita income is three times as high.

Hashim Ali Household, Rampur. Hashim Ali is fifty-two, illiterate, and a *dhobi* by caste. He heads a household of ten members. Of these, three adult women are primarily engaged in household maintenance tasks and dairying, and his three sons are in the work force. The expanding transport sector has provided off-farm work opportunities to several households in Rampur. Two of Hashim Ali's sons are employed as private bus conductors in Meerut, and their earnings constitute by far the largest share of the total household income, which is more than Rs 25,000 per annum; per capita income is more than Rs 2,600. These figures are well above the poverty level and considerably above the average for the landless category in the village. Off-farm activities contribute more than two-thirds of total income. Dairying accounts for another one-fifth share, and the remainder is derived from agricultural labor and crop cultivation on leased land.

Hashim Ali's compound is a blend of the traditional and the modern. Most of the house is *pucca*; however, the walls of the courtyard are of mud painted with dung. Activity centers in the open courtyard, which is shared by livestock, family members, and guests. A hand pump in one corner provides drinking water. A slightly rusted manual fodder-chopping machine stands amidst the debris of chopped *cheri* (sorghum fodder). Two cane chairs and a table, part of his daughter-in-law's dowry, sit beside a sagging *charpoy* (rope cot). Also in evidence on the dirt floor of the courtyard is a sewing machine (another dowry item) with a partially sewn garment. China and stainless steel have replaced the more durable brassware for eating and drinking utensils, but meals are still cooked in the traditional manner, on a *chulha* (clay stove) in the open courtyard, over a fire of dung cakes and dried twigs. Capital assets include five head of livestock, of which two are buffaloes in milk. The cash flow problem is solved primarily by off-farm income which accrues on a monthly basis and is supplemented by selling milk. Food security is partially ensured by earnings in kind from agricultural labor during harvest and sharecropping a small plot of maize. Other food is purchased in the market.

The household's prosperity is fairly recent and is the consequence of the two sons working as bus conductors at a daily wage of Rs 37. After working as apprentices for a year and a half, they were promoted to their present jobs two years ago. Unlike Hashim Ali, who is illiterate, the sons have passed class 8.

Hashim Ali himself works as a washerman for about fifteen households in the village. He is paid in kind for his labor: 20 kilograms of grain and 12 kilograms of *gur* per household during the year. At current prices, the cash equivalent of these in-kind earnings is about Rs 1,600, which is woefully inadequate for subsistence. Hashim Ali's ancestors eked out a living under the old *jajmani* system solely by doing laundry for patron landlords; Hashim Ali, however, engages in his traditional occupation not so much for the direct income it brings as for its indirect benefits, which include access to big farmers' fields for fodder and land for lease. This year he sharecropped 0.3 hectare of maize and received a one-fourth share for his labor input. Big and medium landholders short of labor often lease out a portion of their land and households linked to the landowners through labor contracts receive first priority.

Dairying is an important subsidiary occupation, not only in terms of the income but also as part of the strategy for increasing returns to family labor, especially female labor. In Muslim households there is more division of labor by gender, as cultural taboos prevent women from working outside the home. Women help clean, feed, and milk the buffaloes, and the men cut and carry the fodder. Unlike most other landless households, which sell their milk to the local milk vendor, Hashim Ali sells his to the cooperative dairy, where he receives a higher price. He is able to do this because cash flow is not a binding constraint in his household.

Three years ago Hashim Ali purchased a buffalo with a loan of Rs 4,500 and a one-third subsidy under the IRDP. He already owned one buffalo at the time, but an additional member had been added to the household when his son was married, and dairying provided a useful outlet for the new daughter-in-law's labor. He still has the animal and has repaid one-third of the loan.

Hashim Ali, a modest man, admits that his household has progressed considerably since his father's time, when subsistence was provided solely through rigid caste occupations under a patron-client relationship.

Akhtar Ali Household, Rampur. Akhtar Ali, fifty-two, is illiterate; he is a village *dudhiya*, or milk vendor. Twice a day he collects milk from some twelve households and sells it to a dairy in Meerut, earning a commission in the process. Akhtar Ali is one of two *dudhiyas* in the village. His earnings from this activity constitute more than 80 percent of the total household income. The sale of milk from his own buffalo accounts for another 10 percent, while agricultural labor and sharecropping make up the rest. The total household income is more than Rs 25,000, and per capita income is about Rs 2,900; both are considerably higher than the average for the landless category.

Akhtar Ali heads an extended family of nine members. His two sons, both with only primary education, assist in the milk business and hire out as agricultural laborers during the harvest season. His wife and daughter-in-law manage the household and assist in animal husbandry. The women have no schooling.

The house is of brick with a courtyard. Like Hashim Ali's household, the courtyard is cluttered with evidence of the old and the new. Prominent among the stack of wheat straw and a mound of dung cakes are half a dozen tin cans, the hallmark of the *dudhiya*. Twice a day, rain or shine, Akhtar Ali and his son transport about 80 liters of milk to a private dairy in Meerut in cans fixed to their bicycles. The net daily income is around Rs 60 and takes care of the cash flow goals of the household. The bicycles and the riders show signs of wear; the work is strenuous and adds up to more than four hours of commuting per day. Akhtar Ali plans to purchase a motorcycle to ease the tedium but has not yet been able to save the necessary capital.

The strategy of diversifying into trade with accumulated savings, together with utilization of family labor, has brought considerable prosperity to Akhtar Ali's household. Not content with his present standard of living, Akhtar Ali wants to diversify further by setting up a welding workshop in one of the nearby growth centers. Almost a decade ago Akhtar Ali learned welding, and ever since it has been his heart's desire to go into business for himself. He has been unable to accumulate enough savings from his present activities, though, and capital continues to be his major constraint. He has not applied for institutional credit, having been discouraged by the experience of others who were denied credit by the banks.

Akhtar Ali's ancestors were *telis* (oil pressers) who were also allotted a small plot of land on a sharecropping basis by their patron landlords. With traditional skills and a minimal land base they managed a bare subsistence living. In his youth, Akhtar Ali worked for several years as a permanent servant on the farm of a big farmer. By the late 1960s, however, he felt that agricultural labor had limited potential; the scope of the traditional oil-pressing occupation was dwindling as HYVs of wheat drastically reduced the acreage of *rabi* pulses and oilseeds. He moved to Ahmedabad, where he had relatives, and apprenticed himself to a welder. He worked as a welder in Ahmedabad for six or seven years, remitting most of his earnings to support the family he had left behind in the village.

He returned to Rampur in the late 1970s with the intention of setting up a welding workshop financed by his savings. However, the capital he could raise was insufficient for the purpose, and he set up a milk-vending business instead to take advantage of the booming milk market. Akhtar Ali purchased a couple of bicycles and some 15–liter tin cans and, assisted by his two sons, set up shop in Rampur. At present the household owns two buffaloes and a calf. Although the buffaloes' milk is a supplementary source of income, the animals also represent a form of savings that can be converted readily into cash should an emergency arise.

During harvest season, the son not engaged in the milk business hires himself out as an agricultural laborer. Wages are paid in kind, one-twentieth share of the harvested grain per day, which is equivalent to a cash wage of about Rs 22. The household also sharecrops a small plot of maize, which adds to the food supply.

Shyam Lal Household, Rampur. Shyam Lal, sixty-three, is the head of an extended family of ten members. The principal strategy followed by the household has been to seek upward social and financial mobility through education.

Shyam Lal places a premium on education. His eldest son has a bachelor's degree and is employed in a clerical position in a textile mill. Shyam Lal's household income accrues solely from off-farm activities. His son earns a salary of Rs 1,600 per month at the mill. He uses the bus and *tempo-taxi* to commute more than 20 kilometers to work and prefers to live with the extended family in the village rather than move his wife and children to the factory town.

Income sources consist of the son's salary, which accounts for more than half of the total income, a *parchoon* shop (run by Shyam Lal) that contributes about one-fourth, and the interest on his retirement savings, which has been invested in a time deposit. The total household income is above Rs 30,000—more than twice the average for landless households. Per capita income is about Rs 3,300, also considerably higher than the average.

The household's level of living is reflected in the ownership of consumer durables. The house is a handsome brick structure with a spacious courtyard. The family boasts one of the dozen television sets in the village.

Shyam Lal's ancestors were priests by profession who performed religious rituals in the village. Although the household derived a comfortable living from this, young Shyam Lal saw no future in it. He passed middle school (class 8) and secured employment in a private textile mill more than 20 kilometers from Rampur, to which he commuted every day by bicycle. Today his son makes the same journey, partly by *tempo-taxi* and partly by bus, in half the time. When he started at the lowest tier, Shyam Lal's salary was about Rs 50 per month; when he retired forty years later it had risen to Rs 1,100.

Shyam Lal invested part of his retirement fund, about Rs 5,000, in a *parchoon* shop, which keeps him occupied and brings in an average daily income of around Rs 30. His shop is situated in his house and does a brisk business in items of daily use: cigarettes, sugar, grain, tobacco, salt, spices, onion, tea, cooking oil, and the like. He supplies items on credit and is usually paid on a weekly or monthly basis.

The striking point about Shyam Lal's household is the fact that although not involved in any agricultural activities, the family is domiciled in the village and has in the space of a generation acquired several of the amenities formerly restricted to urban areas.

Sachdeva Mahato Household, Sitapur. Sachdeva Mahato is a dapper, energetic man of fifty-two years who belongs to the Scheduled Caste. He heads an extended family of seven members, including his four sons, a daughter-in-law, and a grandchild.

Sachdeva is shrewd and innovative. Seizing upon the opportunity created by the rapidly increasing demand for modern furniture in villages and rural towns, he and his sons learned the skill of caning chairs. Today this off-farm activity, which they follow during the slack months of the monsoon, contributes almost one-third of the total household income. The demand for chair caning is the direct consequence of increased rural incomes. The business requires minimal investment. The only raw material is nylon cane, which Sachdeva purchases from Meerut. He is usually able to do this with household savings but occasionally needs to borrow from the local milk vendor.

Himself illiterate, Sachdeva places a high value on the formal education of his sons and has encouraged them to acquire manual skills as well. The household has improved its well-being through off-farm diversification into construction labor and caning chairs. In addition, dairying and agricultural labor contribute to family income.

His eldest son recently received his bachelor's degree and is interviewing for various white-collar jobs. A smart young man who dresses in the style of urban middle-class youth, he seems confident of getting a job in the near future. His brothers are still studying and expect to follow in his footsteps. The household's investment in education is thus only beginning to yield returns. Meanwhile Sachdeva's household has managed to elevate its economy above the poverty level. Total income is about Rs 16,000, and per capita income is around Rs 2,300, about par for the average landless household. Off-farm employment contributes more than 60 percent of household income, while dairying and agricultural labor contribute almost equally to the remainder.

The household has come a long way since the time of Sachdeva's father and grandfather. Then income sources were limited to caste occupations and toiling on the landlords' fields in return for a mere pittance. Sachdeva's ancestors skinned dead animals and tanned their hides. They barely subsisted.

The factor that changed the course of Sachdeva's household was his employment in a sugar mill 3 kilometers from his village. For seventeen years Sachdeva was employed as an unskilled worker during the cane-crushing season. His wages enabled his sons to acquire an education. When it became physically difficult for him to lift and carry 100–kilogram bags of sugar on his back, he quit and took up chair caning.

Five years ago Sachdeva purchased a buffalo under the IRDP program, with a loan of Rs 3,000 and 50 percent subsidy. He repaid the loan through sale of milk and then sold the buffalo to meet an important social obligation—his first son's wedding. His present buffalo is the previous animal's progeny. He sells milk to the milk vendor. He plans to enlarge his herd when his second son gets married, but at present there is a shortage of female labor in the household.

Food security, a major goal for the household, is met partially through food earned in kind during the harvest season. Sachdeva and two of his sons hire themselves out as agricultural laborers for about sixty days each year and are paid 10 to 12 kilograms of wheat per day. Their remaining food needs are met through market purchases.

Bashir Mohammad Household, Sitapur. Bashir Mohammad, fifty-two, is a blacksmith by profession. His ancestors made a living by making and repairing agricultural implements for their landlord patrons.

Today Bashir Mohammad has diversified his household economic activities to include an agroprocessing unit; custom hiring of a tractor, thresher, and cart; dairying; and part-time repairing of hand pumps and tubewells. More than 85 percent of the household's income is derived from off-farm activities. Dairying and crop cultivation together make up the rest. Total income is more than Rs 33,000, twice the average for landless households; the per capita figure of more than Rs 6,600 is three times the average. How did this man who started life as a landless farm servant transform himself?

At fifteen Bashir Mohammad left his job as a farm servant and became apprenticed to a blacksmith-carpenter in a town 15 kilometers from the village. Two years later he found wage employment in the sugar mill 3 kilometers from Sitapur. Bashir Mohammad proved mechanically adept and soon learned to repair and maintain such new-technology-related equipment as tubewells, hand pumps, pipes, trolleys, *buggis*, and the like. He saved enough from his wages to purchase the necessary tools and equipment to start his own business. His sons learned the trade and began to assist him. Together they ride their bicycles over country roads in a 10- to 15-kilometer radius to repair taps, hand pumps, and tubewells.

In 1977 Bashir Mohammad, in partnership with two other people, set up the first motor-driven *atta-chakki* (grain mill) in the village. This was possible only because the village had been electrified in the late 1960s. Four years later Bashir acquired sole proprietorship of the mill by buying out his partners. An uninterrupted supply of electricity is still a luxury in the rural areas, however, and the *atta-chakki* operates in alternate day and night shifts.

Bashir also invested in buffaloes with credit from the local *dudhiya* so that the family's considerable female labor force could be effectively utilized. There were at one time nine members in the household; at present there are five. When the number of women was at its peak he had four buffaloes; this is now reduced to two.

As his sons married and moved away, the activities of Bashir Mohammad's household underwent a change. He reduced the dairy and blacksmith operations and replaced them with less labor-intensive activities. In the mid-1980s Bashir applied for a loan to set up a small-scale scissors-manufacturing unit under a government program. He was granted a loan of Rs 15,000 at a subsidized interest rate, but he used it to purchase a secondhand tractor and thresher, which he hires out. His rationale for this change is that between the time of application and the final loan, the son who was to assist him in the venture opted out. Without the son's help the scissors unit could not operate. Bashir has so far tactfully kept bank officials from asking awkward questions. He expects to pay off the loan by hiring out the tractor and thresher.

A younger son is apprenticed to a furniture maker in Delhi and remits Rs 300 per month. Another son manages the *atta-chakki.* Bashir himself works on the tractor and thresher during wheat harvest and hires out the tractor and cart to transport sugarcane during the five months of the cane-crushing season. The household also sharecrops 0.5 hectare of paddy on a one-fourth-share basis, which provides them with their annual rice supply. Paddy cultivation makes efficient use of family labor at a time when its opportunity cost in alternate employment is very low. He still undertakes some repair work on an in-kind payment basis for about twelve big farmers. This satisfies the social obligations needed to enable him to gain access to leased land and fodder for the animals.

Marginal Households

Six case-study households in the marginal landholding category were documented in Rampur and Sitapur villages. Figure 5.9 illustrates the income profile of the two Rampur case studies relative to the average marginal household of the village. Marginal households in the village account for around one-fourth of the total. The annual income of the average marginal household is about Rs 25,000, and the per capita income is more than Rs 3,800. Off-farm activities are the principal source of income and account for nearly 45 percent of annual income; crop cultivation adds about 28 percent.

Dilawar Singh's household exhibits an income profile in which the contribution of off-farm activities is below the average. Total and per capita incomes are also below average. This, however, is a temporary situation: the household is paying off a major loan with which it purchased a *tempo-taxi.* Once the loan is repaid, the household's net earnings will rebound.

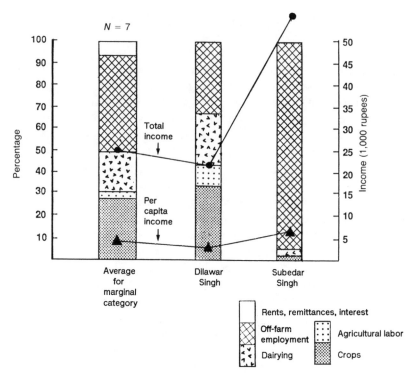

Figure 5.9. Amount and sources of income for marginal case-study households in Rampur, 1988–1989

Subedar Singh's household income is two and a half times the average, and the per capita income is one and a half times higher than the average. The share of income from off-farm sources constitutes more than 95 percent of the household's income.

Figure 5.10 places the four case studies from Sitapur in the village context. Marginal households in Sitapur constitute about one-fifth of the total. The net annual income of the average marginal household is more than Rs 16,000, and the per capita income is about Rs 2,500. A little over 40 percent of the total income accrues from off-farm employment, with crop cultivation accounting for about one-third. All four Sitapur case-study households enjoy incomes higher than the average. The income profiles suggest that higher levels of household income are associated with greater participation in off-farm activities.

Although each of the four households illustrates the importance of

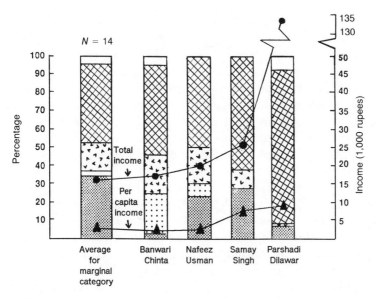

Source: Table A5.11.

Figure 5.10. Amount and sources of income for marginal case-study households in Sitapur, 1988–1989

off-farm diversification as a strategy to raise income, Parshadi Dilawar's household deserves special mention. Parshadi Dilawar is the village milk vendor. His remarkable success in the business has not only led to an upward financial mobility but has resulted in the breakdown of social and cultural taboos. Parshadi belongs to a Backward Caste, yet he and his household are among those leading the economic transformation of the village. He has lent money to several landless and marginal households to enable them to purchase milk cattle. His son is the first doctor from Sitapur.

Dilawar Singh Household, Rampur. At twenty-eight, Dilawar Singh is one of the youngest household heads in the village. His father and uncle jointly cultivated 1.84 hectares. Upon his father's death, Dilawar and his two brothers inherited 0.31 hectare each, which they decided to operate jointly. Dilawar manages the crop cultivation operation assisted by a handicapped brother, while the youngest brother is self-employed in the transport sector; he drives the *tempo-taxi.*

Before settling down to farming, Dilawar Singh tried unsuccessfully to get a job on the strength of his high school diploma. He is one of several young men in the village for whom formal education did not guarantee a job. His brother, the *tempo-taxi* driver, discouraged by Dilawar's experience, dropped out of school after class 8 and went to work in a motor parts factory, where he gained the skills of a mechanic. This served him in good stead four years later when he purchased a second-hand *tempo-taxi*. Presently, the household assets consist of 0.92 hectare of farmland and capital assets in the form of the *tempo-taxi* and four head of livestock. The extended family has ten members, of which three male and three female members make up the labor force.

The household income is around Rs 21,000, and the per capita income is a little above Rs 2,100. Although higher than poverty level they are slightly below average for the marginal landholding category. This situation was likely to be remedied in the course of the year following our study. Earnings from the *tempo-taxi* are Rs 2,100 per month. From this the household is repaying the loan for the *tempo-taxi* with monthly installments of Rs 1,500. Two-thirds of the Rs 30,000 borrowed from a finance company (at an interest rate of 18 percent) has already been repaid and they expect to clear the remaining debt in the next eight to ten months. This will lead to an impressive increase in both total and per capita income.

Dilawar Singh's house is of brick and contains three rooms. One room, used for storage, contains sacks of wheat piled one on top of the other. Plump brown mice scurry about the sacks. Although the house has a connection for power, Dilawar keeps a lantern and kerosene oil handy, for the supply is very erratic.

Dilawar Singh is an average Jat farmer whose major constraint from increasing crop income is the scattered nature of his landholding. His 0.92 hectare of land is fragmented into three plots. For this reason he has not yet invested in his own tubewell and purchases water from various sources. He divides about two-thirds of his gross cropped area between wheat and sugarcane; the remainder is planted to fodder and maize. He intercrops rapeseed with wheat and *urd* with maize. The cropping intensity on his farm is around 225 percent. The wheat and sugarcane yields, at 3.6 and 40 tons per hectare, are on a par with the average in the village. Except for the sugarcane, everything he produces is consumed by the household.

A major change in the household's strategy occurred with the decision to diversify into the transport business. Dilawar's youngest brother was working in a private motor parts factory in Meerut for about Rs 35 per day. Two years ago, he quit his job. The brothers borrowed Rs 30,000 from a private finance company and purchased a secondhand *tempo-taxi*. With his experience in motor parts, Dilawar's brother is able to handle minor repairs and runs the *tempo-taxi* without any problem. He ferries passengers and produce from Rampur to the market town of Dhaulri 10 kilometers away. It is a busy route, and half a dozen *tempo-taxis* do a brisk business. His net daily income is about Rs 70. During the harvest, when family labor is required on the field, Dilawar and his brother share the driving and thereby ensure that agricultural operations are not adversely affected.

Subedar Singh Household, Rampur. Subedar Singh, fifty, is a Jat marginal farmer. His extended family consists of his wife, two sons, their wives, two unmarried daughters, and two grandchildren. Subedar Singh himself has only a primary education; however, his older son has a high school diploma, and the younger has passed class 8. Unlike most other households in the village, both his daughters and the daughter-in-law were also educated up to class 8. Subedar Singh's household embodies the spirit of rural entrepreneurship. Although he started out as a landless agricultural laborer, his family has now become one of the wealthier households in the village.

The household's "industrial" enterprise, a thread-spooling business, fetches a net income of around Rs 4,300 per month and accounts for more than 95 percent of total income. Dairy and crop cultivation account for the remaining 5 percent. The total yearly household income is more than Rs 54,000, twice the average for households in the marginal category, and the per capita figure of around Rs 6,000 is one and a half times the average.

The household's strategy has been to overcome the limitation of a minimal land resource through access to education, skills, and off-farm employment. This family has gone one step further than most in that the most skilled and qualified member has provided employment opportunities for less-qualified siblings. The family objectives have shifted from risk minimization and food security to maximization of returns to capital and family labor. Electricity is a major constraint to further expansion of the enterprise; were this not the case the family's thread-spooling enterprise could provide additional employment in the village.

Subedar Singh's household lives in a handsome brick structure with several rooms. They recently added two rooms and a small shed to house the machines, raw material, and packing boxes. Although the family has many modern conveniences—electric fans, sewing machines, furniture, and the like—a radio and television are conspicuous by their absence. Subedar Singh has a puritanical streak and believes in austerity. On Sunday mornings the younger family members sneak out to a neighbor's house to watch their favorite television show. After much persuasion, Subedar Singh recently allowed his son to purchase a motorcycle. The family is both upwardly and geographically mobile.

It was not always so. Subedar Singh has seen some hard times. His father was a tenant of the *zamindar* of Rampur who cultivated about 3.5 hectares of land on a half-share basis. With the abolition of *zamindari*, he was dispossessed and joined the ranks of the landless agricultural laborers. Subedar Singh himself worked as a permanent farm servant for several years, leasing small plots of land from his employer while his wife toiled as an agricultural laborer and managed two buffaloes. They supplemented their income through the sale of *ghee* (clarified butter) rather than milk. The market for milk was fairly limited at the time; besides, *ghee* could be stored for a longer period. Living a life of bare subsistence, Subedar Singh saw his oldest son through high school. With his qualifications the young man got a job in a mill that made sewing thread and reels. Gradually, through his own enterprise he made his way from messenger boy with a salary of Rs 30 per month to supervisor earning Rs 1,300 per month, all in the span of twelve years. The younger sons did not do as well educationally and passed only class 8. They were unable to get jobs. At this point a major change in the household's strategy took place. In 1985 the oldest son decided to shift from wage employment to self-employment. He had acquired sufficient skill and experience in the sewing thread and reel business to feel confident enough to start his own business. This would simultaneously provide employment to his brothers.

The major constraint was capital. Initially they were denied institutional credit, but they managed to raise the Rs 20,000 for machinery and raw materials from various sources. Prominent among these were accumulated savings, sale of family jewelry, and loans from relatives and a private moneylender who charged an interest rate of 36 percent. The original unit consisted of one machine that wound 200 meters of different-colored sewing thread onto cylindrical cardboard spools, which were then sold as reels of sewing thread.

Marketing of the sewing thread reels, particularly competition with the old established brand names, was a major problem in the first couple of years. The oldest son handled the marketing; in the early years he toured neighboring villages and market towns on his bicycle vending his product while the remaining household members managed the production aspect. When power is available they work the machinery, attending to labeling and packing when the electricity is cut off. Over a period of four years the household paid for the unit. They also established small but assured outlets in the neighboring growth-center town of Dhaulri as well as in Meerut City. In 1988, when the unit was expanded to include more machinery, credit was forthcoming from a commercial bank. By this time the household had established its creditworthiness.

Subedar Singh was landless until the mid-1970s, when he was allotted a plot of 0.17 hectare under the surplus village land distribution program. Needless to say, his income from crop cultivation is insignificant.

Banwari Chinta Household, Sitapur. Banwari Chinta, seventy, was landless until the mid-1970s, when he received 0.07 hectare of land under the surplus land distribution program. His household consists of four adult males, three adult females, and two children. The adults all hire out for agricultural labor. The women are predominantly employed in weeding and harvesting potatoes and vegetables. The household dairy enterprise, comprising four buffaloes, is managed entirely by the women, who even cut and carry the fodder.

Banwari depends on his neighbor's tubewell to irrigate his little plot. The neighbor, a big farmer, is a whimsical man and often refuses to sell water to Banwari, although the latter is willing to pay the water charges. The neighbor often expects Banwari to perform some additional favor for him, such as providing labor on his fields. Banwari often complies with his wishes. A meek man who once worked as a farm servant on the neighbor's land, he prefers to avoid a confrontation. However, conflicts occur when the same demands are made of Banwari's son Uchhu, who does not feel any such obligation. On the contrary, he is an aggressive young man, trying to cast off the shackles of caste subordination. But crop cultivation is not an important activity for Banwari's household; it contributes less than 3 percent of the total income. Off-farm earnings from a small power-driven grain mill contribute half the total, and dairying and agricultural labor account for about one quarter each. Annual income is over Rs 17,000, higher than average for the marginal category.

The strategy followed by Banwari's household has been to diversify income sources and optimize returns to family labor through agricultural labor, dairying, and an agroprocessing unit. The strategy helps reduce risk, provides food security, and has elevated the household economy above the poverty level.

During the cane-crushing season Uchhu works at a *kolhu*. He has been doing this for several years and has acquired expertise in the work. He now commands a daily wage of Rs 40, twice the wage for casual agricultural labor. This has enabled the household to accumulate some savings. One of Banwari's sons is a *jawan* (soldier) in the military. He used to remit regularly about Rs 500. Now his own family commitments allow only an occasional contribution. However, these remittances in the early 1980s made a significant contribution to starting the agroprocessing unit, the third *atta-chakki* in the village. The necessary Rs 18,000 was raised partially from household savings; the rest was borrowed from relatives and from the local moneylender. The motor that operates the grain mill also drives the fodder chopper. Uchhu custom chops fodder for four or five big farm households. The two activities together contribute more than half the household's income. During 1988 the mill was idle for a time as the result of a burned-out transformer. Replacement took more than three months and caused the loss of almost 20 percent of the household's income.

Nafeez Usman Household, Sitapur. Nafeez Usman, sixty, is a *bhisti* by caste and heads a nine-member household. He owns 0.2 hectare of land but has access to an additional hectare for sharecropping paddy and pulses. The sharecropping contract, with a big farmer, is linked to a reciprocal agreement to provide labor at lower than the market wage rate at sugarcane harvest time. Nafeez is also allowed access to fodder for his animals, and his wife helps with the farmer's household chores. It is difficult to quantify who is the gainer. The fact that the system has worked for the past eight years suggests that both benefit.

The household income is around Rs 20,000, higher than average for the marginal landholding category; per capita income at Rs 2,200 is slightly below average. More than half the total income is contributed by off-farm activities, which include rickshaw pulling, a *parchoon* shop, and construction labor. Dairying and crop cultivation each account for a one-fourth share of income.

Household assets consist of a cycle-rickshaw and four head of livestock. The family labor force is made up of four males and three females. Nafeez Usman is illiterate, and his sons have not acquired much formal education either. The oldest son manages the shop, another operates the rickshaw for five to six months in the year and works as a casual construction laborer during the remaining months. Nafeez Usman pulls the rickshaw when he is free of his obligations to the big farmer. A third son is apprenticed to a barber. The dairying operation is managed primarily by the women. The milk is sold to the local milk vendor, from whom small loans are occasionally secured to tide over minor consumption crises such as purchasing gifts for married daughters when they come to visit.

Plying the rickshaw between Sitapur and the nearest growth center, 3 kilometers away, fetches an average daily income of Rs 25. Originally, Nafeez purchased a horse cart under the IRDP. He claims that his horse died and he fell behind on his payments. When he received a recovery warrant, he panicked, sold the cart, borrowed the remainder from the milk vendor, and repaid the loan. He purchased the rickshaw instead and has been operating it for the past two years.

Samay Singh Household, Sitapur. At thirty-nine, Samay Singh is one of the younger household heads in the village. His is a nuclear family of five members, the only other adult being his wife. He dropped out of school after class 8. Ten years ago Samay Singh was part of an extended family and lived with his father and uncle, jointly cultivating more than 5 hectares. In 1979 he fell out with his father and set up a separate household with his share of about 0.5 hectare of land.

The household income sources changed from crop cultivation to off-farm activities. Today, off-farm activities contribute nearly two-thirds of household income, with crop cultivation constituting the remainder. The major off-farm income sources are a power-driven grain mill, a *kolhu*, and custom hiring of farm machinery. The household's net annual income is around Rs 36,000, twice the average for the marginal landholding category; the per capita income, at more than Rs 7,000, is almost three times as high.

Today Samay Singh is a successful farmer and businessman who, within the span of a decade, diversified his income sources on the strength of his entrepreneurial skill. After parting from his extended family, Samay Singh very soon realized that 0.5 hectare of land could not provide the sort of life he had been accustomed to and that he had to diversify into nonfarm activities. His first step was to take a job in a cane-crushing unit for two seasons. This enabled him to accumulate some savings. Together with some capital borrowed from relatives and a major contribution from the sale of his wife's jewelry, these savings enabled him to invest Rs 16,000 in an *atta-chakki*, the second in Sitapur. The grain mill today contributes 30 percent of the household's income.

Samay Singh, a Jat, is an excellent farmer. He owns a tubewell in partnership with three others. His wheat yield, 4.8 tons per hectare, is one of the highest in the village. On his land he plants wheat and fodder. With limited home consumption he is able to sell more than two-thirds of the wheat. A single buffalo, managed by his wife, supplements the household's income. Three years ago, with funds generated by his grain mill and crop cultivation and a loan from a private moneylender (at 24 percent interest), he invested Rs 18,000 in a *kolhu*, a small cane-crushing unit for making *gur*. Today the *kolhu* runs for five months during the crushing season, employs four people, and contributes about 25 percent of the household's income.

In 1987 he diversified further by purchasing, with household savings, a thresher for Rs 6,000. He hires out the thresher during the harvest season. With so many activities claiming his time Samay Singh now has a permanent employee to help run the *atta-chakki*.

Samay Singh is an example of the emerging class of rural entrepreneur. He is a marginal landholder, but intensive crop cultivation and off-farm activities have raised his income well above the poverty line. Hence he does not qualify as a potential beneficiary under the IRDP. Yet he has made major capital investments on three different occasions. Each time he did so partly with his own savings and partly by borrowing from a moneylender at high interest rates. Despite the lack of encouragement from the official agencies Samay Singh not only set up a successful agrobusiness but also provides employment to three or four local young men.

Parshadi Dilawar Household, Sitapur. Parshadi Dilawar, sixty-three, owns less than 0.5 hectare of land and yet is one of the wealthiest men in the village. His household places a high value on education. Parshadi's son is the first doctor from Sitapur village.

The household's total income is more than Rs 133,000, and the per capita figure is around Rs 8,900; both are more comparable with an urban upper-middle-class household than with any rural category. Although the family lives in the village, Parshadi's household has many modern conveniences: a large house, electric fans, television, motorcycle, and other consumer durables.

Parshadi's major occupation, which contributes more than 85 percent of total household income, is the milk business. He is a *dudhiya* with a difference. He does not sell fluid milk, but prepares *mawa* by boiling the milk till all the water has evaporated. The solid *mawa* left behind is the main ingredient in sweets and fetches a price eight times higher than milk. However, the value-adding process is both labor- and capital-intensive and requires considerable fuel. Markets for *mawa* exist in both Meerut and Delhi, and Parshadi sells it there three or four times a week.

A strategy of adding value to product, combined with developing educational qualifications and skills, has paid high dividends for Parshadi Dilawar. It was not always so. Pardshadi's ancestors were landless *kumhars* who made their living under the old patron-client system by baking clay pots. They were also involved in a part-time grain dealership. Parshadi's first job was in a kiln, transporting bricks on donkeys. At sixteen he obtained employment in a sugar factory at a monthly salary of Rs 55. He retired forty-three years later in 1984 at a salary of Rs 1,200.

A reliable source of wage income influenced the economic strategy of the household. The children were encouraged to improve themselves through education. The youngest son graduated as a doctor from the medical college in Meerut, two sons passed high school, and a third learned tailoring.

While holding his job at the factory, Parshadi invested some of his savings in a small *parchoon* shop, managed jointly by himself and his wife. In the mid-1970s he closed the *parchoon* shop and with two grown-up sons set up the milk and *mawa* business. While he was still employed at the sugar factory Parshadi confined his involvement to supervising his sons, but since his retirement he has become more actively involved. Today the milk business has expanded to employ a nephew and four permanent laborers.

Parshadi also keeps a buffalo and a cow and has bought 0.5 hectare of land in the neighboring village to provide cereal and fodder for the household. On retirement he received a retirement fund of about Rs 60,000, part of which he invested in a time deposit. Parshadi has also invested in urban real estate. The son with the tailoring skills intends to diversify into a garment business and set up his shop in the neighboring district. Parshadi is also involved in moneylending, and several landless and near-landless households have been able to acquire livestock with loans obtained from him.

Appendix

Table A5.1. Percentage distribution of landholdings and land owned in Rampur, by landholding group, 1970–1971 and 1980–1981

Landholding group	1970–71		1980–81	
	Number	Area	Number	Area
Marginal (<1 ha)	41	7	32	6
Small (1–2 ha)	16	10	30	20
Medium (2–5 ha)	27	37	28	49
Big (>5 ha)	16	46	10	25
TOTAL	100	100	100	100

Source: Meerut, District Land Records Office, Collectorate.

Table A5.2. Gross cropped area under major crops in Rampur, 1951–1952 through 1988–1989 (ha)

Crop	1951–52	1976–77	1980–81	1988–89
Wheat	34.8	49.0	54.2	56.9
Paddy	—	0.4	9.3	0.1
Maize	18.6	13.8	—	13.0
Pulses	57.5	2.0	3.4	6.4
Oilseeds	—	—	—	0.3
Sugarcane	37.6	81.3	76.1	61.6
Fodder	53.8	40.9	53.4	37.8
Vegetables	0.4	—	2.4	9.3
Fruits	—	—	12.5	17.4
Potato	0.4	—	—	1.1
Others	8.9	5.2	0.7	0.4
TOTAL	212.0	192.6	212.0	204.3

Source: Rampur village land records, 1951–52, 1976–77, 1980–81, and 1988–89, District Land Records Office, Collectorate, Meerut.

Table A5.3. Composition of Rampur work force, 1971 and 1981 (number of workers)

Workers	1971	1981
Agricultural	162	211
	62.3%	57.0%
Nonagricultural	98	159
	37.7%	43.0%
TOTAL	260	370
	100%	100%

Source: India, Office of the Registrar General, *Census of India*, General Population Tables, Series 1 (Delhi: various years).

Table A5.4. Net and gross cropped area per household in Rampur, by landholding group, 1988–1989 (ha)

Landholding group	Net cropped area	Gross cropped area
Marginal	0.61	1.22
Small	1.73	2.49
Medium	2.71	4.48

Source: Rampur household sample survey, 1988–89.

Table A5.5. Percentage distribution of gross cropped area in Rampur, by major crops and landholding group, 1988–1989

Crop	Marginal	Small	Medium
Wheat	32.1	28.9	28.2
Paddy	—	—	—
Maize	5.9	5.5	1.4
Pulses	—	—	—
Oilseeds	—	0.5	2.3
Sugarcane	25.2	24.6	40.4
Fodder	25.9	28.3	23.9
Vegetables	3.5	1.4	—
Fruits	1.5	8.4	3.8
Potato	—	1.4	—
Others	5.9	1.0	—

Source: Rampur household sample survey, 1988–89.

Table A5.6. Annual household income, per capita income, and percentage distribution of income, by source and landholding group for Rampur, 1988–1989 (rupees)

Landholding group	No. of households	Crop income	Noncrop income				Total household income	Per capita income
			Agricultural labor	Dairying	Off-farm employment	Rents, remittances, etc.		
Landless	13	49 (0.3)	2,318 (15.3)	2,111 (14.0)	8,924 (59.2)	1,684 (11.2)	15,086 (100)	2,286
Marginal	7	6,905 (27.8)	614 (2.5)	4,525 (18.2)	10,903 (44.0)	1,864 (7.5)	24,811 (100)	3,877
Small	7	15,551 (68.5)	429 (1.9)	4,986 (22.0)	857 (3.8)	864 (3.8)	22,687 (100)	3,845
Medium	3	28,292 (75.1)	—	5,627 (14.9)	—	3,733 (9.9)	37,652 (100)	5,620
TOTAL	30	8,090 (37.8)	1,248 (5.8)	3,697 (17.3)	6,611 (30.9)	1,740 (8.1)	21,386 (100)	3,342

Source: Rampur household sample survey, 1988–89.
Note: Figures in parentheses are percentages of total household income.

Table A5.7. Cumulative percentage of aggregate household income by source for Rampur, 1988–1989 (N = 30)

Landholding group	Households	Crop income	Noncrop income	Total household income
Landless	43.3	0.3	49.0	30.6
Marginal	66.6	20.2	80.4	57.7
Small	89.9	65.1	92.9	82.5
Medium	100	100	100	100

Source: Rampur household sample survey, 1988–89.

Table A5.8. Annual household income, per capita income, and percentage distribution of income, by source, for landless case-study households in Rampur, 1988–1989

Income per household	Average of landless category		Hashim Ali household		Akhtar Ali household		Shyam Lal household	
	Rupees	Percentage	Rupees	Percentage	Rupees	Percentage	Rupees	Percentage
Crop income (own land)	—	—	—	—	—	—	—	—
Crop income (leased land)	45	0.3	320	1.2	320	1.2	—	—
Agricultural labor	2,308	15.3	2,520	9.7	1,600	6.2	—	—
Dairying	2,112	14.0	5,480	21.1	2,510	9.6	—	—
Off-farm employment	8,931	59.2	17,705	68.0	21,600	83.0	28,700	84.8
Rents, remittances, pensions, interest	1,690	11.2	—	—	—	—	5,148	15.2
Total income per household	15,086	100	26,025	100	26,030	100	33,848	100
Income per capita	2,280		2,603		2,892		3,385	
Main occupation			Off-farm employment		Off-farm employment		Off-farm employment	
Subsidiary occupation			Agricultural labor		Dairying		Pension	

Sources: Rampur household sample survey and Rampur case-study household survey, 1988–89.

172

Table A5.9. Annual household income, per capita income, and percentage distribution of income, by source, for landless case-study households in Sitapur, 1988–1989

Income per household	Average of landless category		Sachdeva Mahato household		Bashir Mohammad household	
	Rupees	Percentage	Rupees	Percentage	Rupees	Percentage
Crop income (own land)	—	—	—	—	—	—
Crop income (leased land)	441	2.8	—	—	500	1.5
Agricultural labor	1,913	12.0	2,880	18.1	—	—
Dairying	3,109	19.5	3,305	20.8	4,400	13.3
Off-farm employment	10,255	64.2	9,700	61.1	24,666	74.4
Rents, remittances, pensions, interest	240	1.5	—	—	3,600	10.9
Total income per household	15,958	100	15,885	100	33,166	100
Income per capita	2,533		2,269		6,633	
Main occupation	Off-farm employment		Off-farm employment		Off-farm employment	
Subsidiary occupation	Dairying		Dairying		Dairying	

Sources: Sitapur household sample survey and Sitapur case-study household survey, 1988–89.

Table A5.10. Annual household income, per capita income, and percentage distribution of income, by source, for marginal case-study households in Rampur, 1988–1989

Income per household	Average of marginal category		Subedar Singh household		Dilawar Singh household	
	Rupees	Percentage	Rupees	Percentage	Rupees	Percentage
Crop income (own land)	6,897	27.8	1,152	2.1	7,206	34.0
Crop income (leased land)	—	—	—	—	—	—
Agricultural labor	620	2.5	—	—	2,000	9.4
Dairying	4,516	18.2	1,500	2.8	4,800	22.6
Off-farm income	10,917	44.0	51,720	95.1	7,200	34.0
Rents, remittances, pensions, interest	1,861	7.5	—	—	—	—
Total income per household	24,811	100	54,372	100	21,206	100
Income per capita	3,860		6,041		2,121	
Main occupation			Off-farm employment		Off-farm employment	
Subsidiary occupation			Dairying		Crop cultivation	

Sources: Rampur household sample survey and Rampur case-study household survey, 1988–89.

Table A5.11. Annual household income, per capita income, and percentage distribution of income, by source, for marginal case-study households in Sitapur, 1988–1989

Income per household	Average of marginal category		Banwari Chinta household		Nafeez Usman household		Parshadi Dilawar household		Samay Singh household	
	Rupees	Percentage	Rupees	Percentage	Rupees	Percentage	Rupees	Percentage	Rupees	Percentage
Crop income (own land)	4,227	25.9	282	1.6	688	3.4	8,087	6.0	9,609	26.9
Crop income (leased land)	1,357	8.3	150	0.9	3,811	19.2	—	—	—	—
Agricultural labor	474	2.9	3,825	22.2	1,200	6.0	—	—	—	—
Dairying	2,358	14.5	3,833	22.2	3,960	20.0	1,722	1.3	3,540	9.9
Off-farm employment	7,203	44.2	8,235	47.8	10,200	51.4	114,111	85.2	22,600	63.3
Rents, remittances, pensions	671	4.1	900	5.2	—	—	9,980	7.5	—	—
Total income per household	16,291	100	17,225	100	19,859	100	133,900	100	35,749	100
Income per capita	2,671		1,914		2,204		8,921		7,150	
Main occupation			Off-farm employment		Off-farm employment		Off-farm employment		Off-farm employment	
Subsidiary occupation			Dairying		Crop cultivation		Crop cultivation		Crop cultivation	

Sources: Sitapur household sample survey and Sitapur case-study household survey, 1988–89.

175

6

Dairying as a Source of Income Diffusion: Izarpur Village

Izarpur village demonstrates how dairying is utilized by small producers to supplement household income. Dairying among landless and near-landless households has gained rapid momentum only since about 1975 as increased rural incomes have enabled households to accumulate savings to invest in buffaloes. The investment process has been aided by government programs that provide cheap credit to the poorer households. The special characteristic of the dairying enterprise is the fact that it is almost entirely managed by women and is usually a subsidiary occupation of the household.

Unlike income from crop cultivation, which favors the big landholders, milk animals and income from dairying are more equally distributed in the village. Earnings from the sale of milk not only improve the level of income of the poorer households but also have a tendency to reduce the income disparities between landholding groups.

Profile of Izarpur Village

As the crow flies, Izarpur is about 18 kilometers from Meerut City. Since it lies about 6 kilometers off the main Meerut-Ghaziabad highway, the actual distance by road is some 23 kilometers. The road leading to Izarpur is paved except for the last kilometer or so, which was scheduled to be paved in 1989–90.

The village settlement is characterized by a mix of *kutcha* and *pucca* houses, with some in the process of transformation. More than a dozen television antennae are visible on the rooftops in the wealthier section of the village, where most of the houses are handsome brick and concrete structures. In the late 1970s and early 1980s the village experienced a boom in construction. The government set up a spinning mill and a cooperative milk-processing factory in Partapur, about 6 kilometers from the village on the main highway.

"Occupants of the Front Room"

Describing the importance of livestock in the household of a western U.P. village in the 1930s, William and Charlotte Wiser wrote: "When the cattle come home in the evening, our village lanes are transformed into stables. . . . After dark, when nights are cold, the animals are led into the cattle room, the front room of every farm house. If the men of the family have no room for themselves apart from the courtyard where the women stay, they sleep on their rope strung cots or beds of straw, among the cattle. On suffocating summer nights, both master and animals sleep in the lane outside the door" (1963:59). While many aspects about the livestock enterprise have changed in the thirty years since the Wisers wrote about Karimpur village near Agra, the importance and pride of place of milk animals can be judged by the fact that they continue to be "occupants of the front room."

Livestock for draft power and as a source of organic manure have lost importance in the production process with the increasing use of mechanization and chemical fertilizers, but they are still the major source of fuel for the poorer households, which use dung cakes almost exclusively for their cooking fires. Bullocks and cows have declined in numbers, but buffaloes have increased. In the altered circumstances, milk animals are increasingly valued as sources of supplementary income.

On an ordinary day it is the dairying activity in the village that catches the eye of the visitor. The four milk vendors begin their day at the crack of dawn, making their rounds to collect their quota of milk. Although there is a cooperative society milk collection center in the village, almost three-fourths of the daily milk production is purchased by the *dudhiyas*, who together collect about 350 liters per day. Two of them transport the milk to dairies and tea shops in Partapur and Meerut City in tin cans fixed to bicycles. Another recently purchased a motorcycle and is therefore able to transport more in half the time.

The fourth uses a horse cart. The *dudhiyas* return by noon, empty milk cans rattling, and the cycle begins again three or four hours later, when the evening milk is collected and delivered.

Women are very visible in the village and in the fields engaged in some aspect or another of dairying. Baking dung cakes and setting them out to dry is a labor-intensive and time-consuming activity. Once dry, these dung cakes are piled in a pyramidlike structure and covered with straw and grass, to be used later as fuel. These *bithauras* can be seen all over the village settlement. Some households sell their surplus *bithauras*. Although there are a couple of bio-gas plants in Izarpur, problems of design and maintenance have prevented this technology from more universal adoption.

The 1981 census recorded a population of 893 persons and 167 households in Izarpur. The preliminary village survey conducted in 1988–89 indicated that the number of households had increased to 200. Of these about 37 percent are landless. Scheduled Caste households predominate in the village, accounting for more than 60 percent of the total. The procedure for selection of sample households is discussed in Chapter 4. The composition of the sample for Izarpur is shown in Table 6.1.

Changes in Agrarian Structure

Land distribution in Izarpur is characterized by a very large number of marginal holdings. The size of the average landholding in Izarpur was 0.61 hectare in 1980–81. The land distribution is highly skewed in the village: small and marginal holdings make up 90 percent of the

Table 6.1. Total households and composition of sample in Izarpur, 1988–1989

Household	Village total		Sample size
	Number	Percentage	
Landless	73	36	11
Landowning	127	63	19
Marginal/near-landless (<1 ha)	100	50	15
Small (1–2 ha)	13	6	2
Medium (2–5 ha)	13	6	2
Big (> 5 ha)	1	0	—
TOTAL	200	100	30

Source: Izarpur preliminary village survey, 1988–89.

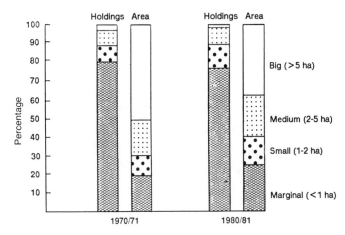

Source: Table A6.1.

Figure 6.1. Percentage distribution of landholdings and land owned in Izarpur, by landholding group, 1970–1971 and 1980–1981

total but operate less than 40 percent of land (Fig. 6.1). The concentration of land, expressed by the Gini coefficient, remained virtually unchanged between 1971 and 1981.

Changes in Employment Patterns

The makeup of the work force in the village points to a growing importance of nonagricultural activities (Fig. 6.2). Of the total number of workers in 1971 about 45 percent earned their livelihood from nonagricultural activities. This had increased to 58 percent by 1981. The 1988–89 household sample survey indicates that more than 80 percent of households now have nonagricultural activities as their main sources of income.

The declining size of the already small landholdings has magnified the need for alternate sources of employment. While a significant number of workers find off-farm employment in Meerut City or in the Partapur growth center, some households have taken up dairying as their major occupation, and almost all households have it as a subsidiary source of income. There is no information on the number of households that have added dairy enterprises in the past decade; however, discussions with farmers indicate that a major increase has taken place since about 1975.

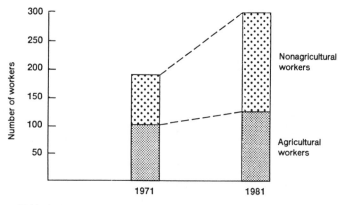

Source: Table A6.2.
Note: Nonagricultural includes animal husbandry, household industry, manufacturing, construction, trade, transport, and services; agricultural includes cultivators and agricultural laborers.

Figure 6.2. Composition of Izarpur work force, 1971 and 1981

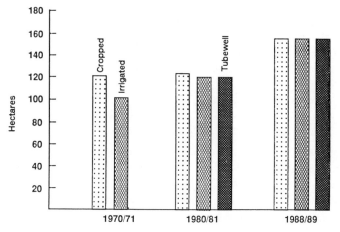

Source: Table A6.3.

Figure 6.3. Net cropped area, net irrigated area, and tubewell irrigated area in Izarpur, 1970–1971, 1980–1981, and 1988–1989

Changes in Agricultural Performance

The impact of the Green Revolution technology is most visible in the transformed nature of irrigation and increases in cropping intensity (Figs. 6.3 and 6.4). In 1970–71 traditional sources accounted for

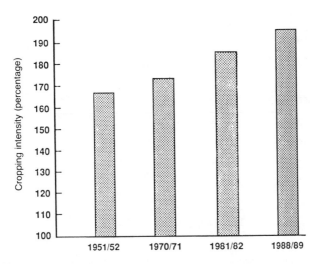

Source: Table A6.4.

Figure 6.4. Cropping intensity in Izarpur, 1951–1952 through 1988–1989

the entire irrigated area. By 1988–89 irrigation was entirely by tube-wells.

Changes in cropping patterns also reflect technological change (Fig. 6.5). The rapid increase in wheat acreage between 1951–52 and 1982–83 was the direct result of the spread of HYV wheat varieties at the expense of coarse cereals, pulses, and oilseeds. The major change in the 1980s was an increase in fodder and sugarcane acreage. The former indicates a rising trend in dairying activity in the village; the latter reflects a tightening of the agricultural labor market.

Figure 6.6 demonstrates that it is the medium landholders who are most affected by the labor shortage. They devote a significant portion of their land to sugarcane, which requires relatively less labor to maintain. The priority given to fodder and cereal crops by the marginal and small landholding groups points to the growing importance of dairying to them.

The Appeal of Dairying to Small Producers

Dairying has always been an important subsidiary activity for land-owning households. The need for draft power and manure made live-stock an important subsystem of the overall farming process. The ani-

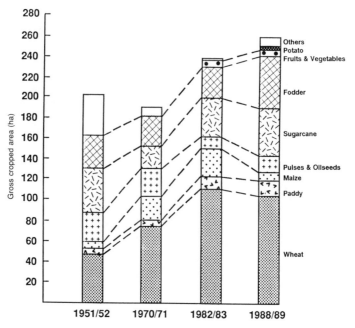

Source: Table A6.5.

Figure 6.5. Gross cropped area under major crops in Izarpur, 1951–1952 through 1988–1989

mal subsystem was in turn sustained by crop residues, which would otherwise have gone to waste. It is easy to understand the desire of farmers to keep milk animals. In addition to providing valuable inputs to cropping, milk animals provide food for the household and supplement family earnings through the sale of dairy products.

The Green Revolution and the associated improvement in rural infrastructure were largely responsible for small producers being able to engage in dairying. The large-scale participation of landless and near-landless households in dairying dates from the late 1970s, a time when major changes were taking place in the rural economy. Agricultural growth was stimulating diversification of the rural economy, and greater numbers of male workers from landless and near-landless households were finding employment in higher-paying off-farm occupations. Private tubewells coupled with the availability of high-yielding varieties of wheat and rice and short-duration varieties of pulses and oilseeds made multiple cropping more common. Multiple crop-

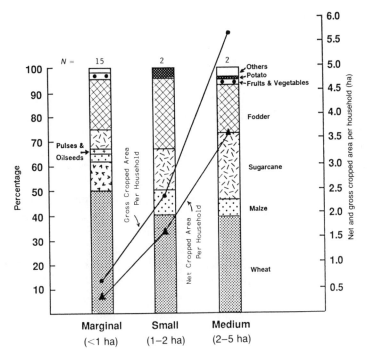

Sources: Tables A6.6 and A6.7.

Figure 6.6. Cropping patterns in Izarpur, by landholding group, 1988–1989

ping, in turn, required tight agricultural schedules, and with male labor increasingly being drawn away into off-farm activities, the bigger farmers were obliged both to mechanize and to shift some of their land into less labor-intensive crops.

Female labor, on the other hand, was becoming increasingly available for additional work. Mechanization had relieved women of such traditional chores as threshing and winnowing, and also from many of their age-old household tasks. Most houses installed hand pumps in their courtyards, relieving the women of the need to draw and carry water from the village well. Rural electrification brought in its wake the *atta-chakki,* which has done much to relieve the village women of the drudgery of manual grinding of wheat. Dairying thus became a logical activity for newly released female labor in the landless and near-landless households. It could be carried out around the home and offered the potential of bringing in considerable revenue.

Participation of Small Producers

More than 80 percent of the Izarpur households in the landless category and more than 67 percent in the near-landless category are involved in dairying. Figure 6.7 indicates the proportion of households in Izarpur that have dairying as their main or subsidiary occupation by size of landholding group. It is clear that more than a quarter of the landless households have dairying as their main occupation, and more than half have it as a subsidiary source of income. Dairying is the major source of income for one-third of marginal households and a subsidiary source for another one-third.

Figure 6.8 shows the percentage distribution of households, milk animals, milk production, milk marketed, and dairy income by landholding category. It is clear that milk animals are fairly equally distributed among the households sampled. In fact, the 37 percent of households that are landless own about one-third of the milk animals, account for the same share of total milk production and income, and contribute 38 percent of all marketed milk. Marginal households, which account for half of all households, contribute nearly half of everything related to dairying.

Milk Marketing. Three channels are utilized for milk disposal in the village. A portion of the milk produced is kept for home consumption. The remainder is sold either to one of the four milk collectors or to the milk cooperative society collection center situated in the village. Until the early 1980s the private milk collectors were the only buyers. With the establishment of a cooperative milk-processing factory at Partapur in 1978, a network of primary milk cooperative societies was established at the village level to purchase surplus milk directly from the producers and transport it to the milk factory. At the milk factory part of the milk is processed into dairy products and the remainder is pasteurized and sold as fresh milk in Meerut and neighboring districts and even in Delhi.

Until the milk cooperative society in Izarpur began functioning, the private vendors held a monopoly in the milk market. The milk vendor does not purchase milk from producers at uniform rates. Since more often than not producers have taken loans or advances from him, the *dudhiya* prices the milk according to the amount and frequency of the loans taken by the producer. The new cooperative milk collection center introduced an element of competition into the village milk market. The price paid by the cooperative society depends on the fat

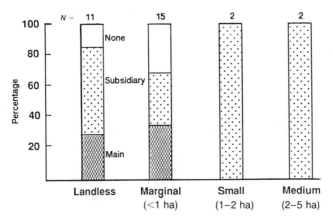

Source: Table A6.8.

Figure 6.7. Percentage distribution of Izarpur households with dairying as the main or subsidiary occupation, by landholding group, 1988–1989

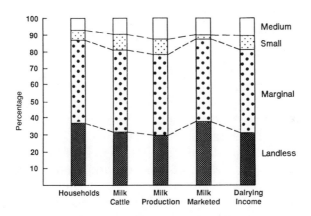

Source: Table A6.9.

Figure 6.8. Percentage distribution of households, milk cattle, milk production, milk marketed, and dairy income in Izarpur, by landholding group, 1988–1989

content of the milk, determined by testing a sample. Payments to producers are usually made on a weekly basis. On average, the price paid by the cooperative society is 15 to 30 percent higher than that paid by private vendors. Despite the higher price, most milk is sold to private collectors because of the credit they extend to small producers. The cooperative society is unable to offer the same service.

Figure 6.9 indicates the pattern of milk disposal by size of land-

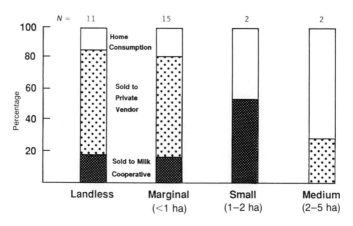

Source: Table A6.10.

Figure 6.9. Disposal of milk in Izarpur, by landholding group, 1988–1989

holding group in 1988–89. As expected, less than one-fifth of the milk produced by the landless group was sold to the cooperative society, about two-thirds went to private milk vendors, and 15 percent was saved for home consumption. Almost the same pattern was followed by the marginal households. All the smallholder households in the sample sold their milk to the cooperative society, and the medium households sold theirs to the milk vendor.

The landless and marginal households generally have a greater need to borrow money for day-to-day expenses or to buy feed for their animals. Hence their greater indebtedness to the milk vendors and the greater volume of milk sold to them. The small landholder households are relatively less constrained by daily cash flow problems and hence are able to benefit from the higher prices offered by the cooperative society. It seems a little unusual that the two medium-sized households in the sample sell their milk to the *dudhiyas* rather than to the cooperative society. One of the reasons given was the savings in labor associated with selling to the milk vendor. The milk vendor comes twice a day to the house and collects the milk, in some cases even milking the animals himself. With the cooperative society one person is required to take the milk to the collection center and wait in line. If the supply of family labor is short, the use of hired labor for this purpose nullifies the gains from the higher price.

The increase in the number of purchasing agencies in the village gives an indication of the extent to which dairying has expanded in

Izarpur. In the late 1970s there were only two private milk collectors operating in Izarpur. By 1988–89 this had grown to four milk vendors and a cooperative society collection center.

Factors Influencing Dairying Activities

The factors most commonly associated with a dynamic dairy industry are availability of credit, availability of surplus female family labor, and availability of fodder. Of these, credit is the crucial factor in starting or expanding an enterprise. Dairy cattle are quite expensive. The average price of a buffalo ranges from Rs 5,000 for one that produces about 5 liters of milk per day to around Rs 10,000 for a 10-liter producer. The role of the government-run IRDP in providing credit for the purchase of buffaloes to landless and near-landless households, which usually do not have access to credit through other channels, has been most favorable.

Credit and the IRDP. The sources of credit available to residents of Izarpur for the purchase of milk animals are (1) institutional credit through commercial banks, which require some form of security or collateral and provide credit at an interest rate of about 14 percent per annum; (2) credit to selected households under the IRDP, which enables the beneficiary household to secure a loan at an 11 percent interest rate, with one-third to one-half the amount of the loan repaid to the bank by the government on behalf of the beneficiary as a subsidy; (3) loans from private moneylenders in Meerut City, whose interest rates range from 24 to 48 percent per annum and who also require as security some form of property, preferably gold jewelry; (4) the local milk vendor, who will provide part of the purchase price of a buffalo provided the milk from the animal is sold to him at below market prices so that he can deduct weekly repayment installments; (5) big and medium farmers in the village who are willing to lend limited amounts to selected households, from whom they expect partial repayment in the form of agricultural labor; and (6) borrowing from relatives and friends, a common practice in Izarpur.

Of the various sources of credit available, the IRDP and local milk vendors have been the most important to expanding the dairy enterprise in Izarpur. The practice of borrowing from big farmers and then working as farm servants to repay the loan, though fairly common in the mid-1970s, seems to have fallen into disuse because the increase in off-farm employment opportunities has led to higher wages in the off-farm sector relative to agricultural labor. Households are reluctant to tie themselves to arrangements that make agricultural labor mandatory. The only credit sources for which collateral is not a condition and interest rates are not prohibitive are the IRDP and the local milk vendor. Under the IRDP the subsidy given to the borrower is paid directly to the bank as partial repayment against the loan. The beneficiary is then required to repay the remaining two-thirds or one-half of the loan, depending on the subsidy, in installments spread over a period of one to three years.

Dairying among the landless and near-landless households in Izarpur has received a favorable boost from the IRDP. Among the sample households of the landless and marginal categories, more than 40 percent received subsidized loans for the purchase of one or two buffaloes during the years 1984–88. Of eleven such households, four had purchased buffaloes within the past two years and therefore had their original assets intact. Four households had purchased their original animal more than two years before and were now managing its progeny. Three other households had also purchased the original animal more than two years ago, had been forced to sell it to meet a family cash emergency, but had since paid off the original loan and acquired a new buffalo through private means.

The IRDP funds are, however, limited; less than one-tenth of all applicants can be selected each year. Other households that want to engage in dairying must arrange their own capital. This is usually done through a combination of household savings, a loan from the milk vendor, and borrowing from relatives. Of these, the institution of the *dudhiya* is the most important. Not only does he extend credit for part of the initial investment, he advances funds for purchasing feeds and concentrates. The popularity of the milk vendor as a credit source also stems from his willingness to provide small amounts of consumption credit.

It is difficult to obtain an accurate estimate of the amount and frequency of borrowing from private sources, as households consider this confidential and are reluctant to divulge information on the subject. But among the households sampled in Izarpur, probably at least 50 percent of the landless and near-landless households that operate a dairy enterprise had availed themselves of credit from the local milk vendor in some form or the other between 1986 and 1989.

Availability of Female Family Labor. Availability of surplus female family labor is an important determinant of the scope and extent of the household dairy enterprise. A widowed woman head of the family often maintains her household by keeping one or two buffaloes. An all-male household, on the other hand, rarely has any milk cattle. Changes in the size of dairy enterprises occur when the number of females in the household increases or decreases. This happens when sons marry and bring home new brides or when daughters marry and move away.

Dairying is eminently suited for women now released from their traditional agricultural tasks and from the drudgery of carrying water and grinding grain; it fits in neatly with their household and child-rearing activities. In most landless and near-landless households, the entire dairying operation—cutting and carrying the fodder; chaffing and preparing the feed; feeding, cleaning, and milking the animals; sweeping and cleaning the animal shed or courtyard; keeping account of the daily transactions; and collecting dung and baking dung cakes—is done by women. In Muslim households, where women are culturally restrained from working outside the home, men cut and carry the fodder, but otherwise women perform all tasks.

Availability of Fodder. Even with the availability of credit and female family labor, lack of fodder can limit the size and scope of a household dairy enterprise. Although concentrates can be purchased in the market, green fodder, which forms the bulk of animal feed, must be found within the village. Near-landless households plant fodder on their small plots of land during the *kharif* season; during the *rabi* season they must plant wheat for subsistence. Both landless and near-landless households also ensure fodder availability by working as hired agricultural laborers. This enables them to procure wheat straw as payment in kind during the *rabi* harvest. During the winter months, from November to February, harvesting sugarcane leads to in-kind payment in the form of *agolas* (sugarcane tops), which are used as animal feed. During the monsoon months of July to October, when natural vegetation is abundant, grasses and weeds cut and gathered from the roadside, fallow fields, or the village commons are fed to the animals. This is supplemented to a small extent by the purchase of standing fields of fodder, usually maize or sorghum in the *kharif* season and *berseem* in the *rabi* season.

The existing cropping patterns are such that landless and near-landless households are able to maintain, on an average, about two buffaloes and two calves per household. Fodder is the ultimate constraint to the size of their enterprises.

Income Diffusion in Izarpur

The income profile of Izarpur points to considerable diversification of the village economy. Crop cultivation does not constitute the major share of village income. Off-farm employment and dairying are more important sources. Since off-farm employment and dairying favor the landless and near-landless, their increasing importance in the total income suggests an improvement in the income levels of the poorer households.

Income from Dairying

Figure 6.10 shows the level and composition of household income for the various landholding groups. The household income of the medium landholder category is about four times higher than that of the landless household, while per capita income is only about two and a half times higher. The main reason for this is household size; in the case of landless households it is about 5.7 persons, whereas for medium households it is more than 9 persons.

Dairying is an important source of income for the landless and near-landless, accounting for about one-fourth and one-third of the total income, respectively, for these two categories. For small and medium households its share is relatively lower.

Off-farm employment provides the largest share of income for both the landless and marginal categories. For the small and medium categories, crop cultivation and income from rents account for the major share. The medium category of landholders took advantage of the demand for housing generated in the villages around the developing industrial estate of Partapur in the late 1970s and early 1980s. A majority of the migrant laborers who work in the government-run spinning mill live in the cluster of villages within a 5- to 10-kilometer radius of Partapur. The demand for housing increased rapidly in the decade of the 1980s, and many medium-level farmers saw an opportunity to invest their farming profits into new rental dwellings. A medium-level farmer may now have as many as five or ten tenants on his property, with rents making a substantial contribution to total household income.

Comparison with the Poverty Line. A comparison with the poverty line of Rs 8,600 per household of five members revealed that of the total sample ($N = 30$), only three, or 10 percent, had incomes below

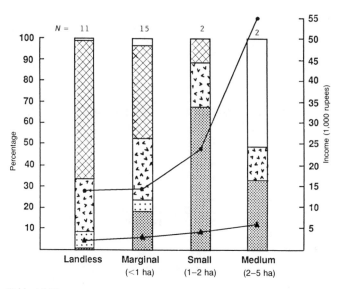

Source: Table A6.11.

Figure 6.10. Amount and sources of income in Izarpur, by landholding group, 1988–1989

the poverty line in 1988–89. Of these, two were landless and one belonged to the marginal landholding category.

Implications for Income Distribution

Figure 6.11 shows the Lorenz curves and Gini coefficients for crop income, dairying income, and total income in Izarpur. It is clear that crop income is the most unequally distributed. Dairying income, on the other hand, has a more equal distribution.

The results of the decomposition of the Gini coefficient by source are shown in Table 6.2. Column 2 indicates that off-farm earnings

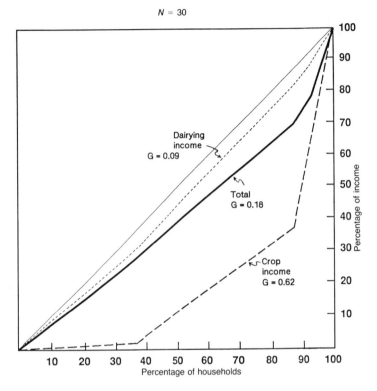

N = 30

Source: Table A6.12.

Figure 6.11. Lorenz curves for Izarpur, by landholding group, 1988–1989

constitute the major share of village income. Dairying also plays an important role, accounting for almost one-fourth of the total. Income from crop cultivation comes in a poor third.

The Gini coefficient (G) in column 3 captures the distribution of each income source independent of the other sources. Crop income by itself is very unequally distributed (G = 0.62), but there is a decline in total village income inequality when the impact of other sources is taken together (from 0.62 to 0.18). The Gini coefficients for individual income sources indicate that dairying income has the most equal distribution, while crop and rent incomes have the least equal distribution.

Column 4 of Table 6.2 presents the correlations between each income category and total income. The importance of the correlation is

Table 6.2. Decomposition of income inequality for Izarpur, by income source, 1988–1989

Income source	Income share (S)	Gini of source (G)	Correlation with rank of total income (R)	Share of inequality		Elasticity of total inequality by income source
				SGR	Percentage	
(1)	(2)	(3)	(4)	(5)	(6)	(7)
Crop income	0.21	0.62	1.00	0.13	70.9	0.50
Agricultural labor	0.04	0.20	−0.77	−0.01	−3.8	−0.08
Dairying	0.24	0.09	1.00	0.02	11.7	−0.12
Off-farm employment	0.37	0.24	−0.84	−0.07	−40.5	−0.77
Rents, remittances, pensions, interest	0.14	0.82	1.00	0.11	61.6	0.48
TOTAL	1.00	0.18	1.00	0.18	100	—

Source: Izarpur household sample survey, 1988–89.

evident when one compares the contribution of each income source to village income inequality (column 6). Crop income and income from rents account for a large part of total inequality (71 and 62 percent, respectively), whereas dairying makes up a relatively small percentage. Agricultural labor and off-farm employment tend to counteract the inequality caused by crop and rent income.

The elasticity of inequality is shown in column 7. A 1 percent increase in crop and rent income for all groups of households would increase income inequality by 0.50 and 0.48 percent, respectively, and a similar change in agricultural labor, dairying, and off-farm employment would reduce inequality.

Household Strategies

In Chapter 5 we discuss some of the broad objectives of the poorer households and the strategies they adopted to improve their economic status. Here we merely repeat that most landless and near-landless households have very limited access to land and capital. Their major resource is family labor. Highest priority is therefore given to identifying and exploiting supplemental income-earning opportunities. Family welfare objectives include food security, minimization of risk, ensuring an adequate cash flow, and the fulfillment of social and cultural obligations.

The dairying activity in Izarpur is well suited to these objectives. Although the output per animal is low, the returns to labor are high because surplus female labor is used. The opportunity cost of female labor is quite low. This is especially true when labor-intensive crops such as vegetables and potatoes do not play a significant role in the cropping patterns of a village, as is the case in Izarpur.

The strategy adopted by landless and near-landless households in starting or expanding a dairy enterprise consists of procuring capital for the purchase of a milk animal and then using female family labor to manage the enterprise. Dairying is usually the subsidiary occupation of the family, the major one being the one in which the men of the household are involved: either off-farm employment or crop cultivation. Access to capital is accomplished through household savings, loans from private sources, or through the IRDP. The enterprise expands and contracts with the demographic cycle of the household, expanding as women in the family increase, contracting when they leave home.

Case Studies

In the following sections this income-generating strategy is illustrated through individual case studies.

Landless Households

The seventy-three landless households in Izarpur constitute 37 percent of the village's two hundred households. They are proportionately represented in the random sample. The average size of the landless household is 5.7 persons. Figure 6.12 indicates that their annual income averages about Rs 13,500, and the per capita figure is about Rs 2,400. Agricultural labor accounts on the average for 7 percent of total income, dairying for 26 percent, and off-farm employment for about 65 percent.

All three households included in the case studies exhibit a per capita income slightly higher than the average. The Dharmi Mansingh household has an income composition fairly similar to the average,

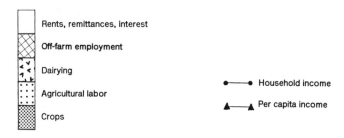

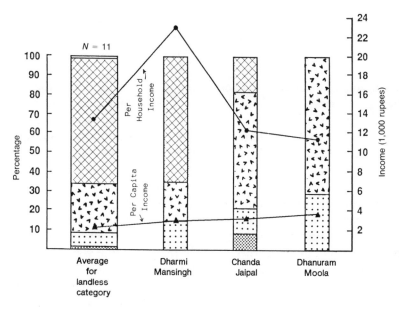

Source: Table A6.13.

Figure 6.12. Amount and sources of income for landless case-study households in Izarpur, 1988–1989

while the Chanda Jaipal and Dhanuram Moola households derive a much higher proportion from dairying.

Dharmi Mansingh Household, Izarpur. Dharmi Mansingh is forty-five, illiterate, and from the Scheduled Caste. He heads a household of eight members. His father, wife, and daughter tend to the family dairy enterprise. The eldest of his two sons finished high school and now manages the family *parchoon* shop, which is situated in one room of the house. The younger son, who dropped out of school after class 5, hires out as a casual laborer for both agriculture and off-farm work. Dharmi Mansingh himself is employed on a seasonal basis, six to seven months a year, as an unskilled laborer in a soft-drink factory about 10 kilometers away. He commutes to work on his bicycle, although the family recently purchased a motorcycle. His is one of the well-off households in the landless category.

The income profile of the household is fairly similar to that of the average for the landless category. The chief source of revenue is off-farm employment and includes earnings from wage employment in the factory plus profits from the *parchoon* shop. Dairying is an important source of subsidiary income and contributes more than one-fifth of the total, which at Rs 23,000 is considerably above the average.

Dharmi Mansingh is halfway through converting his house from *kutcha* to brick. Two rooms have brick walls and a concrete roof; a third room, which opens onto the village lane and houses the small shop, has also been converted into a *pucca* structure. The front room, which serves as a cattle shed in bad weather, still has mud walls and a thatched roof. Dharmi Mansingh expects to convert it to brick within the year.

The courtyard is the hub of activity. For most of the day and the major part of the year the animals are tethered here. Dharmi's wife feeds and waters the animals twice a day and collects their dung, which she piles in one corner. Later in the afternoon, when the chores are done, she and her daughter squat in front of the dungheap and make cakes, which they plaster onto the inside of the courtyard wall to dry. Once dried, they are either used right away for fuel or stored under hay.

Dharmi Mansingh has a hand pump in his yard for drinking water. The house does not have electricity, although the village does. The need for one family member to forgo several days of labor to process the necessary paperwork has so far inhibited the family from applying for a connection. They are quite content to use the kerosene lantern in the evenings. The family usually retires early and is up at dawn. Consumer durables in the household include a radio, two bicycles, a couple of wristwatches, a sewing machine, and assorted pieces of furniture. A secondhand motorcycle purchased recently, one of only half a dozen in the village, has added status to the family.

The success story of Dharmi Mansingh can be attributed to two factors. One is the seasonal job he has held for close to fifteen years. This fairly regular source of income, supplemented by employment as a casual daily laborer, permitted the family to build up some savings. The other factor is the IRDP program, of which Dharmi has taken uncommon advantage. The family received subsidized loans under the program on three different occasions, twice for the purchase of a buffalo and once to set up the *parchoon* shop. The first loan was obtained in 1982 and the last in 1987. The buffalo purchased in 1982 was sold after three years; the first loan was also repaid around this time. The second loan could be procured only after the first was repaid; however, Dharmi managed to get the third loan before he had repaid the second. It is not clear how he managed to arrange for the rules to be bent.

The dairying enterprise consists of one buffalo in milk and two female calves. The present buffalo is the progeny of the second buffalo (purchased under the IRDP), which was sold when its milk yield began to decline. The buffalo yields about 5 liters of milk per day, which is sold to the cooperative society collection center. Fodder for the buffalo is obtained as in-kind payment for agricultural labor. During the winter months all adult members of the household harvest sugarcane for the big and medium landholders. Wages are paid in the form of *agolas*, which are used as animal feed. During the wheat harvest payment in kind allows them to accumulate wheat straw for their animals. Dharmi purchases straw during the harvest season and stocks it for use later in the year, and fields of *cheri* (jowar fodder), which last from July through September. This is supplemented by grass and weeds from the roadside and from field borders. Concentrates such as oil cake and wheat bran are purchased from the market.

There is no doubt that the dairying enterprise and the shop—both financed through the IRDP—have added greatly to the income and status of the household. It is, however, doubtful whether this household should have qualified as a beneficiary for the second and third times. It was not possible solely through the family's recall to obtain an accurate picture of their finances three to four years ago, especially since Dharmi is very reluctant to provide information on such a sensitive point, but it appears likely that the household's income was well above the poverty line when the second and third subsidized loans were received. Be that as it may, the present fairly high income level of the household is largely a result of government aid under the IRDP.

Chanda Jaipal Household, Izarpur. Chanda Jaipal is thirty-six years old, illiterate, and from the Scheduled Caste. His wife is the only other adult member in his family. He has two sons; a daughter died in infancy. His is a fairly average household, but it was selected for study because dairying is the main occupation; as Figure 6.12 shows, more than 60 percent of household's income derives from dairying.

The household's objectives are to ensure food security and maximize returns to family labor. To achieve this Chanda Jaipal hires himself out as an agricultural laborer during the wheat and paddy harvests and is paid in kind, which supplies the household with wheat and paddy straw and sugarcane tops. During agriculturally lean periods he works as a construction laborer. Chanda Jaipal and his wife participate equally in the dairying enterprise, with Chanda cutting and carrying the fodder while his wife chops it, feeds the animals, and takes care of all other tasks. They have two buffaloes, one cow, and a calf. One buffalo and a cow were in milk at the time of our study. The buffalo yields an average of 6 liters per day, the cow 3 liters per day. The milk is sold partly to the cooperative society and partly to the local milk vendor. Chanda Jaipal borrows frequently from the milk vendor to purchase concentrates for the animals and occasionally to tide him over family crises. Although he can get a higher price for the milk at the cooperative, he prefers to maintain a link with the milk vendor.

The household owns no land, but Chanda has access to two small plots. On one he sharecrops paddy on a one-fourth share. For another plot of 0.05 hectare he pays an annual rent of Rs 300. On this, maize fodder and *cheri* are planted in the *kharif* season, and *berseem* in the *rabi*. The family labor resources consist of Chanda and his wife, and their only capital asset is their livestock. Their dwelling is still a mud structure, although a stack of bricks and a couple of bags of cement indicate Chanda's intention to start some brick construction. There is no domestic power connection, but music from a transistor radio floats in the air. A bicycle leans in one corner of the small courtyard; the household also has a sewing machine.

But it is the livestock that seem to fill the entire space in the courtyard. Indeed, the economy of the household centers on the animals. In 1984 Chanda separated from his father and set up his own home. At the time his only source of income was from agricultural labor and off-farm construction work. He applied for a loan and subsidy for the purchase of a buffalo under the IRDP. Being from the Scheduled

Caste he was entitled to a 50 percent subsidy on his loan. His wife managed the buffalo while he continued to supplement household income with wage labor. They repaid the loan in about two and a half years. One of his present buffaloes is the progeny of the original. Two years ago he purchased another buffalo, this time partly with savings and partly with a loan from the milk vendor. He is still paying off the loan. The cow was given to him by his father-in-law when his second son was born in 1986. The major constraint to increasing the present number of animals is a shortage of family labor. He expects to expand the dairy enterprise after another couple of years when his children are older and can help with the chores.

In order to diversify his household operation further, and also to acquire a skill, Chanda Jaipal enrolled in a TRYSEM program to learn tailoring. The training lasted six months, and the participants were provided with loans to purchase sewing machines. Although Chanda Jaipal did buy a sewing machine, he is an indifferent tailor and does not have the confidence to work professionally.

Chanda Jaipal's relative success has been largely due to his dairy enterprise. Since his is a small nuclear family and his children are still quite small, it was not possible to maximize family labor resources through employment outside the house. Hence the strategy was to start an enterprise that would exploit female labor within the household. In this effort the timely assistance from the IRDP was helpful. He was subsequently able to expand the operation on his own.

Dhanuram Moola Household, Izarpur. Dhanuram Moola is twenty-two years old and one of the youngest household heads in the village. The only other adult family member is his wife. He dropped out of school after class 7 and has been working as a casual agricultural laborer. In 1986 he moved away from his father's house and set up his own home. The following year he started a dairy enterprise, and this has added significantly to his household income.

Figure 6.12 shows that dairying contributes more than 70 percent of household income, the remaining being earnings from agricultural labor. The annual income of Rs 11,300 is slightly lower than the category average, but the per capita figure, about Rs 3,800, is considerably higher. That is to be expected since there are only three members in the family.

Until 1988 the only resource of the household was family labor. Both Dhanuram and his wife hired out as agricultural laborers, and Dhanuram also worked as a construction laborer. Today Dhanuram has two buffaloes and two calves. His mud house consists of two rooms and a courtyard. There are not many consumer durables in evidence, which is expected since the family is very young and just starting out on its own. Dhanuram does have an old bicycle. He dresses like an urban youth and wears a smart watch—a wedding gift from his father-in-law.

In 1987 he applied for and secured an IRDP loan and subsidy for the purchase of two buffaloes. In the past, beneficiaries were given loans to purchase one buffalo at a time; a second loan was permitted only after the first was repaid. In 1987 the regulations were changed, and beneficiaries are now given the choice of purchasing either one or two buffaloes; accordingly they now receive either Rs 4,500 or Rs 9,000. The maximum permissible subsidy is Rs 3,000.

Dhanuram applied for and was granted a loan for Rs 9,000 and a subsidy of Rs 3,000. He purchased two buffaloes and two female calves. The milk yield per buffalo is about 5 liters per day. All milk is sold to the milk vendor, who had provided Dhanuram with a loan to purchase concentrate and wheat straw. Dhanuram and his wife continue to work as agricultural laborers in order to procure green fodder for their animals. During wheat, paddy, and sugarcane harvests they are paid in kind. Dhanuram has not repaid any part of his IRDP loan. He believes that the loans are going to be written off in the future (some political parties offered this as bait during the 1989 election campaign).

Dhanuram is contemplating the purchase of yet another buffalo using savings from his dairy business and a loan from relatives. The dairy enterprise has improved not only his household economy but also his status in the village. A number of young men want to emulate his example.

According to Dhanuram, his father received an IRDP loan and subsidy and purchased a buffalo several years before Dhanuram purchased his. With his father's help and the assistance of the newly elected village *pradhan*, a relative, his own loan application was processed within two months. Dhanuram complained that he received only two-thirds of his subsidy amount; the remainder was considered the "hidden cost" of the transaction. Notwithstanding his complaints about this, Dhanuram Moola has no doubt that his present improved economic and social positions are largely due to the loan and subsidy he received under the IRDP.

Marginal Households

Of the two hundred households in Izarpur village, half are in the marginal landholding category. The average size of marginal households is about 5.5 persons. The total and per capita incomes of the average household in this category, as read on the right-hand scale of Figure 6.13, are about Rs 14,400 and Rs 2,600, respectively. Crop cultivation accounts for less than 20 percent of total income for the average household, dairying contributes about 30 percent, and off-

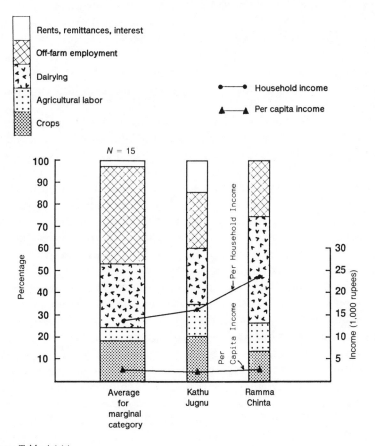

Source: Table A6.14.

Figure 6.13. Amount and sources of income for marginal case-study households in Izarpur, 1988–1989

farm activities make up the major share, about 45 percent. Other sources play a relatively minor role. The two case-study households have per capita incomes almost identical to the average but somewhat higher total revenues. In the Kathu Jugnu household dairying is a subsidiary occupation; for the Ramma Chinta household it constitutes the major source of income.

Kathu Jugnu Household, Izarpur. Kathu Jugnu, forty years old, is illiterate and belongs to the Scheduled Caste. He heads a nuclear household of eight members. Under land reform measures he received 0.3 hectare of village common land in the mid-1970s. In the early 1980s, with a loan and subsidy obtained under the IRDP, he installed a pumpset and boring for irrigation. Ever since he has been double cropping his land, planting wheat in *rabi* and paddy in *kharif*. In addition to crop cultivation Kathu Jugnu also hires himself out as a mason.

Kathu's eighteen-year-old son dropped out of school after class 4 and is apprenticed to a truck driver. His second son, sixteen years old and also a dropout, is learning to be a garage mechanic. The fourteen-year-old daughter helps her mother with the household chores and the household dairy enterprise. Crop income constitutes about one-fifth of total household income, and dairying and off-farm employment each account for a one-fourth share. Agricultural labor and renting out of the pumpset contribute another 15 percent each.

The family strategy includes diversification of income sources and full utilization of family labor. Kathu Jugnu hires out as a construction laborer and works on crop cultivation. His wife manages the dairy enterprise and works as a hired agricultural laborer during the wheat and paddy harvest seasons. The children help with the dairying and work as agricultural laborers. Agricultural labor enables the family to stock up on both foodgrains for themselves and on wheat and rice straw for their animals; the wages they receive are paid in kind.

The present dairy enterprise consists of one buffalo and a calf. Kathu Jugnu purchased his first buffalo in 1984 and has been maintaining one buffalo ever since. In the early 1980s, after the purchase of a pumpset under the IRDP, he began double cropping his land and also started renting out his pumpset. This improved the family's finances. In the late 1970s and early 1980s the development of the industrial estate in Partapur led to a boom in construction, and Kathu Jugnu used his earning power as a mason to the utmost. The family was able to save enough from his earnings to pay off the government loan and start a dairy enterprise without having to borrow. The milk is sold to the local vendor, from whom they procure short-term loans to purchase concentrates and meet family emergencies.

Kathu has converted part of his house into *pucca* construction. He does not officially have a domestic power connection, but an electric stove among the household items suggests that power somehow finds its way off the transmission line. There is a manually operated fodder cutter in one corner of the courtyard. The family draws water from its own hand pump. Consumer durables include a bicycle, a radio, and a wristwatch.

Kathu expects that by 1991 his older son, as a truck driver, and the younger, as a garage mechanic, will be able to supplement the household's income handsomely. He plans for the older son to marry within the next two years. When that son's wife joins the family Kathu hopes to expand his stock of buffaloes to two. He would like to diversify further by setting up a power-driven oil-pressing machine. The acreage of oilseeds in the neighboring villages is increasing, and this encouraged him to make enquiries about the enterprise. His major constraints are lack of credit and the uncertain power supply.

Ramma Chinta Household, Izarpur. Ramma Chinta, forty-five, is illiterate and belongs to the Scheduled Caste. He owns about 0.3 hectare of land, part of which was allotted to him from surplus village land under the government's land reform program. His family consists of seven members: himself, his wife, a married son, a daughter-in-law, a daughter, and two younger children. His older son dropped out of school after class 8 and joined the work force. The fifteen-year-old son is now studying in class 9.

The household's strategy is diversification. The men engage in off-farm wage employment chiefly as masons in construction work, sometimes in the village but more often in Partapur (6 km away) or Meerut City. During harvest season all members of the family work as agricultural laborers and are paid in kind. In this way they are able to earn about one-third of their annual wheat and rice requirements. On their own plot of land they cultivate wheat in the *rabi* season and fodder in the *kharif*. Nearly half of the household income comes from dairying. Off-farm employment and agricultural labor contribute 25 and 13 percent, respectively.

Dairying graduated from a subsidiary to the main family occupation in 1988 when Ramma Chinta added two more buffaloes to his existing stock of two buffaloes and a cow, motivated by the addition of a new daughter-in-law to the family. His existing stock had been slowly built up over the previous six or seven years using savings and loans from relatives and the local milk vendor. At one time he pledged his wife's silver jewelry to raise Rs 500 from a moneylender in Meerut at 36 percent interest.

Encouraged by the example of several other households who obtained subsidized loans under the IRDP, he applied for a loan of Rs 9,000 and a subsidy of Rs 3,000 to purchase two buffaloes. It took four to five months for the loan to be approved. Ramma Chinta was not satisfied with the quality of buffaloes that could be purchased with Rs 9000, so he borrowed a further Rs 4,000 from the milk vendor and bought two buffaloes for Rs 13,000. The average milk yield of these buffaloes is about 8 liters per day. During the study period his four buffaloes were in various stages of lactation and their yields varied from 4 to 8 liters per day.

Ramma Chinta sells his entire milk output to the local milk vendor. In this way he is already repaying the loan obtained from him. He is in no hurry to repay the government loan, having heard from various quarters that there is a possibility that these loans may be written off at some future date.

Ramma Chinta's newfound prosperity is evident in a new table fan and a couple of new chairs set out in the courtyard. The house itself, of brick, is freshly whitewashed. It is, however, his healthy-looking buffaloes standing beside the feeding trough that catch the eye. Almost all household activity seems to revolve around them.

Ramma Chinta confided that being the owner of four buffaloes had improved not only his financial position but also his social status in his community and in the village. One measure of his improved status is that he was able to arrange a better match for his daughter than would have been the case previously.

Appendix

Table A6.1. Percentage distribution of landholdings and land owned in Izarpur, by landholding group, 1970–1971 and 1980–1981

Landholding group	1970–71		1980–81	
	Number	Area	Number	Area
Marginal (<1 ha)	79	19	76	25
Small (1–2 ha)	9	11	12	15
Medium (2–5 ha)	9	19	10	23
Big (>5 ha)	3	51	2	37
TOTAL	100	100	100	100

Source: Izarpur village land records, 1970–71, 1980–81, District Land Records Office, Collectorate, Meerut.

Table A6.2. Composition of Izarpur work force, 1971 and 1981 (number of workers)

Workers	1971	1981
Agricultural	102 54.8%	123 41.8%
Nonagricultural	84 45.2%	171 58.2%
TOTAL	186 100%	294 100%

Source: India, Office of the Registrar General, *Census of India*, General Population Tables, Series 1 (Delhi, various years).

Table A6.3. Net cropped, net irrigated, and tubewell irrigated area in Izarpur, 1970–1971, 1980–1981, and 1988–1989 (ha)

Area	1970–71	1980–81	1988–89
Net cropped area	123	125	155
Net irrigated area	102	120	155
Area irrigated by tubewells	—	120	155

Source: Izarpur village land records, 1970–71, 1980–81, 1988–89, District Land Records Office, Collectorate, Meerut.

Table A6.4. Cropping intensity in Izarpur, 1951–1952 through 1988–1989

Area	1951–52	1970–71	1981–82	1988–89
Net cropped area (ha)	139.0	123.0	125.0	155.4
Gross cropped area (ha)	232.2	213.7	232.7	305.3
Cropping intensity (percentage)	167	174	186	196

Source: Izarpur village land records, 1951–52, 1970–71, 1981–82, 1988–89, District Land Records Office, Collectorate, Meerut.

Table A6.5. Gross cropped area under major crops in Izarpur, 1951–1952 through 1988–1989 (ha)

Crop	1951–52	1970–71	1982–83	1988–89
Wheat	47.4	73.7	110.5	104.3
Paddy	6.1	5.3	22.1	13.1
Maize	4.9	23.9	28.3	11.9
Pulses	34.0	27.1	11.7	6.5
Oilseeds	0.4	—	—	7.2
Sugarcane	40.9	22.3	38.5	53.1
Fodder	30.8	32.8	20.4	44.5
Vegetables	0.4	0.8	0.4	0.9
Fruits	—	—	5.6	5.8
Potato	0.4	0.4	0.4	1.9
Others	37.0	5.1	0.1	10.9
TOTAL	202.3	191.4	238.0	260.1

Source: Izarpur village land records, 1951–52, 1970–71, 1982–83, 1988–89, District Land Records Office, Collectorate, Meerut.

Table A6.6. Net and gross cropped area per household in Izarpur, by landholding group, 1988–1989 (ha)

Landholding group	Net cropped area	Gross cropped area
Marginal	0.39	0.63
Small	1.64	2.42
Medium	3.70	5.67

Source: Izarpur household sample survey, 1988–89.

Table A6.7. Percentage distribution of gross cropped area, by major crops and landholding group for Izarpur, 1988–1989

Crop	Marginal	Small	Medium
Wheat	50.3	40.0	38.5
Paddy	11.3	—	—
Maize	3.8	10.4	6.7
Pulses	2.1	—	—
Oilseeds	—	—	—
Sugarcane	8.0	17.3	26.6
Fodder	20.7	28.6	19.7
Vegetables	0.9	—	—
Fruits	1.3	—	2.2
Potato	—	3.5	1.1
Others	1.6	0.2	5.2

Source: Izarpur household sample survey, 1988–89.

Table A6.8. Izarpur households having dairying as main or subsidiary occupation, by landholding group, 1988–1989

Landholding group	Main occupation	Subsidiary occupation	No dairy enterprise	Total number of households
Landless	3	6	2	11
	27.2%	54.5%	18.2%	100%
Marginal	5	5	5	15
	33.3%	33.3%	33.3%	100%
Small	—	2	—	2
		100%		100%
Medium	—	2	—	2
		100%		100%

Source: Izarpur household sample survey, 1988–89.

Table A6.9. Percentage distribution of households, milk cattle, milk production, milk marketed, and dairying income in Izarpur, by landholding group, 1988–1989

Landholding group	Households	Milk cattle	Milk production	Milk marketed	Dairying income
Landless	36.6	31.7	30.0	38.3	30.9
Marginal	50.0	49.0	48.3	48.6	49.8
Small	6.7	9.7	8.5	2.7	7.6
Medium	6.7	9.7	13.2	10.3	11.6
TOTAL	100	100	100	100	100

Source: Izarpur household sample survey, 1988–89.

Table A6.10. Disposal of milk in Izarpur, by landholding group, 1988–1989

Landholding group	Consumed by household	Sold to private milk vendor	Sold to cooperative milk society
Landless	15.0%	66.3%	18.7%
Marginal	20.0%	64.0%	16.0%
Small	48.0%	—	52.0%
Medium	72.0%	28.0%	—

Source: Izarpur household sample survey, 1988–89.

Table A6.11. Annual household income, per capita income, and percentage distribution of income in Izarpur, by source and landholding group, 1988–1989 (rupees)

Landholding group	No. of households	Crop income	Noncrop income				Total household income	Per capita income
			Agricultural labor	Dairying	Off-farm employment	Rents, remittances, etc.		
Landless	11	160 (1.2)	936 (6.9)	3,515 (26.0)	8,760 (64.8)	141 (1.0)	13,513 (100)	2,355
Marginal	15	2,622 (18.2)	833 (5.8)	4,154 (28.9)	6,325 (44.0)	440 (3.1)	14,374 (100)	2,598
Small	2	15,687 (67.4)	—	4,785 (20.6)	—	2,804 (12.0)	23,276 (100)	4,232
Medium	2	18,703 (34.0)	—	7,264 (13.2)	—	29,000 (52.7)	54,967 (100)	6,107
TOTAL	30	3,662 (21.1)	760 (4.4)	4,169 (24.0)	6,374 (36.7)	2,392 (13.8)	17,357 (100)	2,976

Source: Izarpur household sample survey, 1988–89.
Note: Figures in parentheses indicate percentage of total household income.

Table A6.12. Cumulative percentage of aggregate household income for Izarpur, by source, 1988–1989 ($N = 30$)

Landholding group	Households	Crop income	Dairying	Total household income
Landless	36.6	1.6	30.9	28.5
Marginal	86.6	37.4	80.7	69.9
Small	93.3	66.0	88.4	78.9
Medium	100	100	100	100

Source: Izarpur household sample survey, 1988–89.

Table A6.13. Annual household income, per capita income, and percentage distribution of income, by source, for landless case-study households in Izarpur, 1988–1989

Income per household	Average of landless category		Dharmi Mansingh household		Chanda Jaipal household		Dhanuram Moola household	
	Rupees	Percentage	Rupees	Percentage	Rupees	Percentage	Rupees	Percentage
Crop income (leased land)	162	1.2	—	—	955	7.7	—	—
Agricultural labor	932	6.9	3,200	13.9	1,600	12.8	3,200	28.4
Dairying	3,513	26.0	4,800	20.9	7,500	60.2	8,085	71.6
Off-farm income	8,771	64.9	15,000	65.2	2,400	19.3	—	—
Rents, remittances, pensions, interest	135	1.0	—	—	—	—	—	—
Total income per household	13,513	100	23,000	100	12,455	100	11,285	100
Income per capita	2,359		2,875		3,114		3,762	
Main occupation			Dairying		Off-farm employment		Dairying	
Subsidiary occupation			Off-farm employment		Dairying		Agricultural labor	

Sources: Izarpur household sample survey and Izarpur case-study household survey, 1988–89.

Table A6.14. Annual household income, per capita income, and percentage distribution of income, by source, for marginal case-study households in Izarpur, 1988–1989

Income per household	Average of marginal category		Kathu Jugnu household		Ramma Chinta household	
	Rupees	Percentage	Rupees	Percentage	Rupees	Percentage
Crop income (own land)	2,587	18.0	3,254	19.6	1,788	9.6
Crop income (leased land)	29	0.2	—	—	540	2.9
Agricultural labor	834	5.8	2,520	15.2	2,500	13.4
Dairying	4,154	28.9	4,150	25.0	8,875	47.7
Off-farm employment	6,324	44.0	4,200	25.3	4,900	26.3
Rents, remittances, pensions, interest	446	3.1	2,500	15.0	—	—
Total income per household	14,374	100	16,624	100	18,603	100
Income per capita	2,598	—	2,078	—	2,658	—
Main occupation			Off-farm employment		Dairying	
Subsidiary occupation			Dairying		Off-farm employment	

Sources: Izarpur household sample survey and Izarpur case-study household survey, 1988–89.

7

Labor-intensive Cultivation of High-Value Crops: Jamalpur Village

Jamalpur village demonstrates labor-intensive cultivation of high-value crops as a mechanism of income diffusion. The cropping patterns of the village underwent major changes during the 1980s. The HYVs of wheat introduced in the mid-1960s led to dramatic increases in wheat yields as well as shifts in cropping patterns in favor of wheat acreage. In the 1980s, however, yield increases slowed and there was no further increase in wheat acreage. Wheat profitability began to decline, and farmers started to shift their land into such high-value crops as potatoes and vegetables. This coincided with the rise in both rural and urban incomes and the increased demand for nonfoodgrain crops.

The cultivation of high-value crops is largely restricted to the big and medium landholders, and most poorer households have gained only indirectly through the additional opportunities created for agricultural laborers. Marginal landholders face several constraints in the widespread cultivation of such crops. However, several marginal landholders have devised strategies to overcome the constraints of land, capital, and assured irrigation and now produce the crops themselves.

Profile of Jamalpur Village

Jamalpur is somewhat different from the villages we have examined thus far. The economic diversification in evidence in the other villages

211

is not so apparent in Jamalpur, which depends rather more on crop cultivation.

The village lies about 38 kilometers south of Meerut City, 5 kilometers off the Meerut-Hapur highway (Fig. 3.1). The nearest growth center, Hapur, in the adjoining district of Ghaziabad, is about 10 kilometers away. Jamalpur is located in Kharkhauda Development Block, one of five blocks where the potato is emerging as an alternate cash crop. During the past decade the annual growth rate of the acreage under potatoes in Kharkhauda Block grew by about 10 percent annually. However, the share of potatoes in the total gross cropped area is still a modest 5 percent. Potato cultivation has acquired greater importance in some villages in the area, and Jamalpur is one of them. About 15 percent of the total cultivated area in Jamalpur is planted to this crop.

A bus rider during the *rabi* season on the Meerut-Hapur highway passes potato fields on both sides of the road. One can count as many as five cold-storage facilities on the final 30-kilometer stretch between Meerut City and Jamalpur. Most of them have been established in the past ten years. During February and March the fields are dotted with men and women harvesting potatoes, and the roads become crowded with carts, tractor-trolleys, and trucks transporting potatoes to market yards, the railway station, or cold-storage facilities.

To reach the Jamalpur *abadi* from the main highway people usually hire the horse cart, which is operated by a village resident. The village *abadi* does not exude the air of economic diversity found in the other villages. There are no motorcycles, and few television antennae are visible atop roofs. Although the wealthier section of the village has a cluster of handsome brick houses, in the poorer section of the village most houses are still *kutcha*. Only the main lane in the village is paved with bricks; the side lanes are still dirt tracks.

Surrounding the *abadi* area of the village are very small plots of land planted to vegetables and enclosed by thorn fences to keep out stray cattle. Narrow paths lead from the village to the fields beyond. The fields all lie within 2 kilometers of the *abadi*. Consolidation was completed in 1978–79, and most small cultivators have a single compact plot of land.

The 1981 census recorded a population of 578 persons and 104 households in the village. The preliminary village survey conducted in 1988–89 indicates that the number of households had increased to 145. Table 7.1 shows the total households in the village and the sample composition.

Table 7.1. Total households and composition of sample for Jamalpur, 1988–1989

Households	Village total		Sample size
	Number	Percentage	
Landless	29	20	6
Landowning	116	80	24
Marginal/near-landless (<1ha)	72	50	15
Small (1–2 ha)	29	20	6
Medium (2–5 ha)	10	7	2
Big (>5 ha)	5	3	1
TOTAL	145	100	30

Source: Jamalpur preliminary village survey, 1988–89.

Changes in Agrarian Structure

The average Jamalpur landholding in 1981 was 1.34 hectares, but the actual land distribution in the village was and is very unequal. Figure 7.1 indicates that in 1981 big and medium landholdings were less than 20 percent of the total but owned more than 60 percent of the land. Marginal landholdings, on the other hand, made up more than 60 percent of all holdings but only 15 percent of the total area.

Although tenancy is not documented anywhere and all contracts are verbal, we estimate that about 23 percent of the land is leased out.

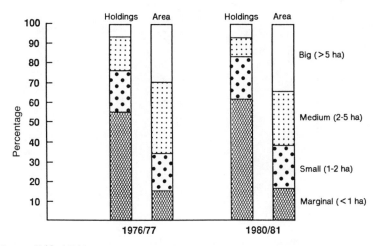

Source: Table A7.1.

Figure 7.1. Percentage distribution of landholdings and land owned in Jamalpur, by landholding group, 1976–1977 and 1980–1981

The most common arrangement is sharecropping, usually for a labor-intensive crop. But sometimes landlords—usually teachers and lawyers who cannot devote much time to farming—lease their land on an annual basis.

Changes in Employment Patterns

The village was almost wholly agricultural in 1971, with 97 percent of the labor force engaged in agriculture. Some diversification occurred during the 1970s, and by 1981 this figure had declined to 75 percent (Fig. 7.2). In 1988–89, 60 percent of the households sampled reported agriculture to be their major source of income.

Changes in Cropping Patterns

Figure 7.3 illustrates the changes in cropping patterns that took place in the village between the early 1950s and 1988–89. The most striking feature is the increase in gross cropped area; it almost doubled during the period. Also evident is a changing attitude toward wheat. The area planted to it doubled with the introduction of the HYVs in the mid-1960s but exhibited no further increase in the

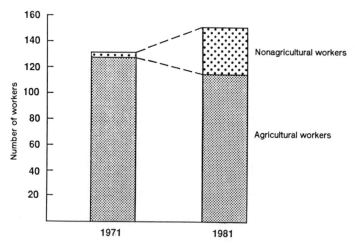

Source: Table A7.2.
Note: Nonagricultural includes animal husbandry, household industry, manufacturing, construction, trade, transport, and services; agricultural includes cultivators and agricultural laborers.

Figure 7.2. Composition of Jamalpur work force, 1971 and 1981

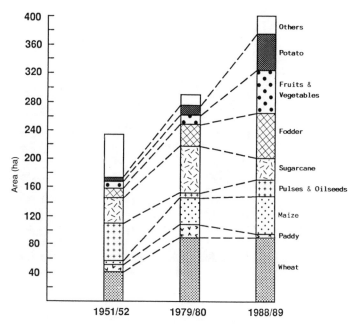

Source: Table A7.3.

Figure 7.3. Gross cropped area under major crops in Jamalpur, 1951–1952, 1979–1980, and 1988–1989

1980s. The impressive increase in the area planted to potatoes, vegetables, and fruits, which more than quadrupled between 1979–80 and 1988–89, points to a new diversification of cropping patterns in Jamalpur. Some of this increase has come at the expense of sugarcane.

The key factor underlying the doubling of the gross cropped area was a change in the nature of irrigation (Fig. 7.4 and 7.5). Whereas tubewells accounted for only half the net irrigated area in the early 1970s, by 1988–89 the entire area was watered with tubewells. The sampled households reported that in 1988–89 all big, medium, and small landholders owned tubewells, as did half of the marginal households; the other half purchased water from neighboring tubewells. In 1988–89 there were forty private tubewells in the village, with an average command area of 4.9 hectares per unit.

It is not possible to compare the yields of the major crops over time. However, farmers reported that the main increases in wheat yields occurred during the first phase of the Green Revolution and plateaued at about 3.5 tons per hectare. The farmers also indicated that the profitability of wheat, which increased sharply during the late

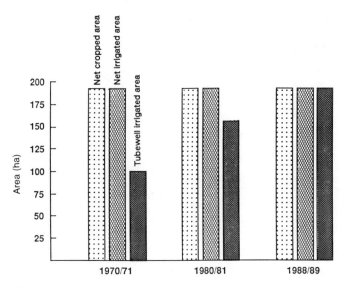

Source: Table A7.4.

Figure 7.4. Net cropped area, net irrigated area, and tubewell irrigated area in Jamalpur, 1970–1971, 1980–1983, and 1988–1989

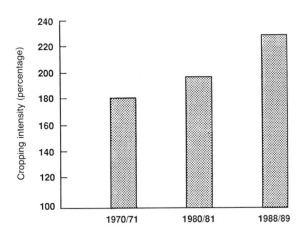

Source: Table A7.5.

Figure 7.5. Jamalpur cropping intensity, 1970–1971, 1980–1981, and 1988–1989

1960s, has in recent years declined relative to other crops. This is the reason why many farmers, especially medium and big landholders, have begun to look for alternate cash crops for the *rabi* season. One such crop is potatoes; winter vegetables are another.

Growing Importance of High-Value, Labor-intensive Crops

Two factors have led to the increase in acreage of high-value, labor-intensive crops: the declining profitability of wheat and the growing demand for fruits, vegetables, potatoes, and other crops as incomes rise and diets become more varied. The data presented in Table 4.2 suggest that the income elasticity of demand in India for vegetables, including potatoes, is about 0.70 for both the rural and urban populations, well above that for cereals and other staple foods.

Appeal of High-Value, Labor-intensive Crops

Table 7.2 shows the major crops of Jamalpur, their cropping seasons, net returns per unit of land, labor requirements per hectare, and returns per labor-day. The net returns to land and family labor (crop income) were estimated after deducting operating costs from the gross value of output.[1] The net returns to land and family labor from potato and vegetable cultivation range from one and a half to two times higher than from wheat. They are about equal to the net returns from sugarcane. But sugarcane locks up the land for a whole year, whereas potatoes and vegetables can be cultivated in a two- or three-crop sequences.

A comparison of labor requirements by crop shows that pulses are the least labor-intensive. The varieties of pulses sown are of short duration; they mature in about sixty days and are generally managed

Table 7.2. Returns per hectare and man-day of labor input for Jamalpur, by major crop

Crop	Crop duration (months) J A S O N D J F M A M J	Net returns per hectare (rupees)	Man-days per hectare	Returns per man-day (rupees)
	◄— kharif —►◄— rabi —►◄— zaid —►			
Wheat		5,545	83	67
Maize		2,880	72	40
Pulses		2,100	45	47
Sugarcane		9,900	166	60
Fodder		3,570	—	—
Potato		9,570	205	46
Vegetables	or or	7,740	261	30

Source: Jamalpur household sample survey, 1988–89.

1. Crop income estimation is discussed in detail in Chapter 4.

by family labor. Wheat cultivation is becoming less labor-intensive. Hiring tractors for field preparation is quite common even among the marginal and small landholders. The use of mechanical threshers is now universal, and scarcely any wheat threshing is done manually. Sugarcane cultivation requires about 166 labor days per hectare. However, since the crop remains in the field for a year, the labor requirement is best compared with a sequence of two or three other crops. The field preparation for sugarcane is now mechanized. Harvesting is still a manual operation staggered over two to three months and usually performed by female laborers who are paid in kind with sugarcane tops.

The labor requirement for potatoes is almost two and a half times that for wheat, primarily because cultivation of potatoes is still largely manual. The sowing and harvesting is usually done by men. During the harvest men dig up the tubers and leave them on the field, and women pick up the potatoes and pile them in a stack. Grading and bagging are also generally carried out by women, and weighing and standardizing of bags is done by men.

Vegetable cultivation is the most labor-intensive of all. It is difficult to estimate accurately the labor requirement for vegetables because they are grown in overlapping sequences and require multiple pickings. The operations are usually carried on by female family labor and vary with different vegetables and households. Our respondents were typically male household heads who were unable to recall accurately the number of days their womenfolk spent in the cultivation of various vegetables. However, we got the impression that the labor requirement for vegetable cultivation is almost three times that of wheat.

Since this labor is largely performed by women whose opportunity cost is close to zero, the returns per labor-day—though only Rs 30—represent an important net gain to the community. Figure 7.6 shows the number of labor-days of hired agricultural workers per household in the four study villages. Because of potato and vegetable cultivation, the number of labor-days per household in Jamalpur is considerably higher than in the other villages we have discussed.

Changes in Cropping Patterns

The shifts in cropping patterns observed at the village level do not apply uniformly to all landholding groups. An examination of cropping patterns by farm size indicates that the big and medium land-

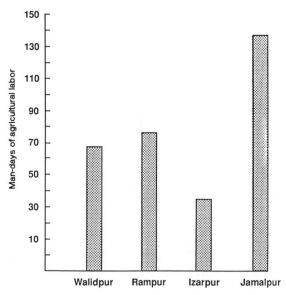

Source: Table A7.6.

Figure 7.6. Man-days of hired agricultural labor per household, by village, 1988–1989

holders have been able to diversify more than members of the small and marginal groups (Fig. 7.7).

It is clear that the medium farmers best reflect the changes in cropping patterns in the village in the past decade. The most striking feature is the share of potatoes and vegetables, which account for more than 65 percent of total cultivated area for this group of farmers. Although vegetables and fruits are shown together, the medium class of landholders has no acreage under fruit. It is the medium farmers who are responsible for generating the additional agricultural labor demand.

The big landholder household, on the other hand, has shifted one-third of its cultivated land to potatoes. The fruit and vegetable acreage, less than 10 percent of the total, is devoted entirely to mango orchards. Orchard fruits, although they provide high returns, are not labor-intensive. In this landholding category it is only potato cultivation that creates agricultural employment opportunities.

In comparison with the medium and big landholders, the marginal and small cultivators plant a much smaller proportion of gross cropped area to potatoes and vegetables. Both marginal and small

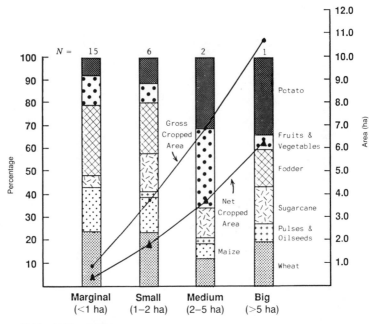

Figure 7.7. Cropping patterns, by landholding group, in Jamalpur, 1988–1989

landholders, especially the former, face several constraints in the more extensive cultivation of these high-value, labor-intensive cash crops. However, in recent years a few households have devised strategies to overcome some of these limitations, and their acreage under these crops has increased.

Obstacles to Adoption by Marginal Landholders

Several factors inhibit marginal cultivators from devoting a larger proportion of their land to potato and vegetable cultivation. The most constraining factor is their lack of land. But a lack of assured irrigation, insufficient capital, and a shortage of storage all act to discourage them as well.

Land Constraint. The average size of the marginal landholdings in Jamalpur is less than 0.5 hectare. Since ensuring food security is the foremost concern of these households, cereals occupy a large share of the land, and since dairying is an important subsidiary occupation for many households, part of the land is planted to fodder. This leaves little room for high-value crops.

Some marginal landholders augment their limited land base by leasing land, usually under a sharecropping arrangement. This explains the relatively high incidence of tenancy in Jamalpur as compared with the other villages we studied. Land is most often leased for potato and vegetable cultivation. Although sharecropping contracts vary in terms of cost-sharing, commonly the entire labor input is provided by the sharecropper. The output is generally shared equally between the owner and the sharecropper.

Competition with Wheat. A field sown to potatoes usually cannot be sown to wheat. The potato is a *rabi* crop generally sown in October and harvested in March. For many marginal landholders, wheat in the *rabi* season is essential for subsistence. Some households, however, have devised a strategy to cultivate both potatoes and wheat on the same plot of land by sowing potatoes in early October and harvesting in early December. Although this practice yields about half a regular crop, it enables the cultivator to plant a late variety of wheat. The lower potato yield is somewhat offset by the higher off-season price the potatoes fetch. Very few households operate in this manner. Those that do typically have a large number of potential workers so that timely operations can be carried out. They also lack significant sources of off-farm income.

Lack of Assured Irrigation. Both potato and vegetable cultivation require large and controlled amounts of water. About half of the marginal landholders do not own tubewells. In 1988–89 the cost of installing an electric tubewell of 5-horsepower capacity, with a command of about 3 hectares, was more than Rs 20,000—about one and a half times the annual income of the average marginal household. It is virtually impossible for such a household to invest in a tubewell. The general practice is to purchase water from a neighbor's tubewell. In the summer months, when the water table falls and the power supply becomes more erratic, the tubewell owner sometimes does not have surplus water to sell or will raise his charge by requiring the marginal cultivator to provide labor for his fields.

Some households have overcome this constraint by purchasing tubewells in partnership, others by entering into a sharecropping contract with a medium or big landholder who already owns a tubewell. Under the latter arrangement the landlord provides the irrigation, the sharecropper contributes the labor, and other costs are shared.

Limited Access to Capital. Another reason potatoes do not find favor with marginal cultivators is the high cost of production. Seed accounts for half the total cost. Unlike wheat seed, which is usually saved from the previous year's production, potatoes are perishable and cannot be stored at home. Each year fresh seed must be purchased from the market.

Cooperative credit for cash crops is available to society members at the rate of about Rs 1,000 per hectare. This represented only a fifth of the cost of potato seed in 1988–89 and about a tenth of the total cost of production. Cooperative credit is not available for the potato-sowing season (October) if the cultivator has already taken out a co-operative loan for a *kharif* crop that will be harvested in November.

To overcome these problems, marginal landholders usually do not cultivate potatoes on their own land, but do so only if they are able to obtain land on lease. Then sharecropping contracts are adjusted so that the cost of seed is partly met by the landlord.

Problems of Marketing. Potatoes and vegetables are perishable commodities. Unlike wheat and maize they cannot be stored at home to be disposed of during the lean season for higher prices. In any case most marginal households cannot afford to wait for higher prices during the off-season; they need immediate cash to meet consumption, debt, and other social obligations. In fact, many near-landless households prefer to grow vegetables rather than potatoes because they are not harvested all at once and daily sales can help ease the family's day-to-day cash flow problems.

For those who cultivate potatoes there are two marketing channels: immediate sale in a market where prices during harvest season invariably fall below support prices, and storage in a cold-storage facility with sales coming later. The cold-storage capacity in Meerut District can now accommodate over 85 percent of production. This has improved the overall potato market and enabled marginal producers to obtain better prices for their produce.

Table 7.3. Amount and sources of income for landless and marginal case-study households in Walidpur, Rampur, Izarpur, and Jamalpur, 1988–1989

| | Landless | | | | Marginal | | | |
| | Per capita income (Rs) | Percentage of total income | | | Per capita income (Rs) | Percentage of total income | | |
Village		Crop income	Agricultural income	Total of crop and agricultural labor income		Crop income	Agricultural income	Total of crop and agricultural labor income
Walidpur	2,533	2.8	12.0	14.8	2,671	34.3	2.9	37.2
Rampur	2,286	0.3	15.3	15.6	3,877	27.8	2.5	30.3
Izarpur	2,359	1.2	6.9	8.1	2,598	18.2	5.8	24.0
Jamalpur	2,048	3.0	24.3	27.3	2,525	35.6	19.3	54.9

Sources: Tables A3.9, A3.10, A6.11, and A7.9.

223

Income Diffusion in Jamalpur

The shift in cropping patterns in favor of potatoes and vegetable is reflected in a higher contribution of agricultural labor and crop income to the revenues of the households in the landless and marginal categories in Jamalpur as compared with the other study villages (Table 7.3).

The share of agricultural labor income for the landless and marginal categories, at 24 and 19 percent, respectively, is far higher in Jamalpur than in any of the other villages.

Figure 7.8 shows the level and composition of household income for the various landholding groups. The household income of the big landholder category is more than six times that of the landless households, and the per capita income is about three times greater. This is a rather larger spread than noted in the other sample villages, but in these villages no big landholders were sampled.

Off-farm employment is the most important source of income for the landless households, followed by agricultural labor. It is noteworthy that agricultural labor accounts for one-fourth of household income. In the case of households in the marginal category, crop income accounts for more than one-third of the total, while agricultural labor constitutes one-fifth. For the small, medium, and big categories, crop income plays by far the most important role.

A comparison with the poverty line of Rs 8,600 per household of five members reveals that of the total sample ($N = 30$), four households, or 13 percent, had incomes below the poverty line in 1988–89. Of these, two were landless and two were in the marginal landholder category.

Implications for Income Distribution

The Lorenz curves and Gini coefficients for crop income, noncrop income, and total income for Jamalpur are shown in Figure 7.9. The unequal distribution of crop income is evident. The results of the decomposition of the Gini coefficient are shown in Table 7.4 and indicate that of the constituents of noncrop income, agricultural labor income and off-farm employment have a negative correlation with total income and therefore tend to counter the inequality of crop income.

The impact of small changes in income from various sources on

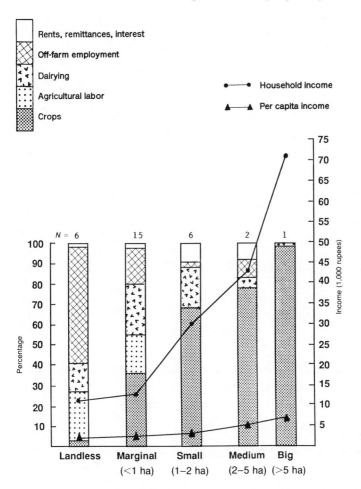

Source: Table A7.9.

Figure 7.8. Amount and sources of income in Jamalpur, by landholding group, 1988–1989

village income distribution (column 7) indicates elasticity of total inequality. In Jamalpur a 1 percent increase in crop income would result in a rise in inequality by 0.51 percent. A similar change in agricultural labor, dairying, and off-farm incomes would lead to a reduction in inequality.

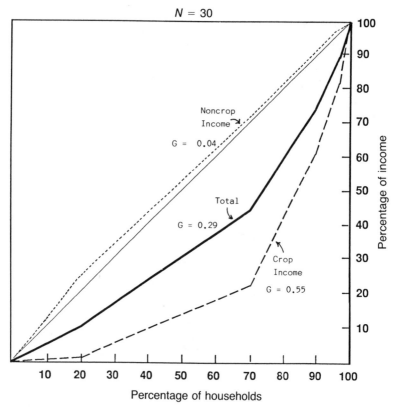

Source: Table A7.10.

Figure 7.9. Lorenz curves for Jamalpur, by landholding group, 1988–1989

Household Strategies

Cultivation of high-value, labor-intensive crops has enabled the landless and near-landless to exploit their one sure resource—an abundance of family labor—to improve their welfare. The landless do this simply by offering themselves as workers to farmers who culti-vate potatoes and vegetables. The near-landless do this also, but some among them have managed to become producers by surmounting the constraints of land, capital, and irrigation.

The land impediment is typically overcome by obtaining land on rental or on a sharecropping basis. Almost one-fourth of the land in Jamalpur is sharecropped. Since most sharecropping contracts call for cost-sharing, this arrangement also helps the marginal farmers meet

Table 7.4. Decomposition of income inequality in Jamalpur, by income source, 1988–1989

Income source	Income share (S)	Gini of source (G)	Correlation with rank of total income (R)	Share of inequality SGR	Share of inequality Percentage	Elasticity of total inequality by income source
(1)	(2)	(3)	(4)	(5)	(6)	(7)
Crop income	0.55	0.57	1.00	0.31	105.7	0.51
Agricultural labor	0.09	0.42	–0.73	–0.03	–9.5	–0.19
Dairying	0.17	0.24	0.68	0.03	9.4	–0.08
Off-farm employment	0.14	0.30	–0.91	–0.04	–13.4	–0.28
Rents, remittances, pensions	0.05	0.41	1.00	0.02	7.8	0.03
TOTAL	1.00	0.29	1.00	0.29	100	—

Source: Jamalpur household sample survey, 1988–89.

227

any initial costs of cultivation. Another strategy is to release land hitherto planted in staple foods for planting to high-value cash crops by using off-farm sources of income to purchase food. Such sources can also help with the purchase of farm inputs.

Even with land available, the absence of assured irrigation can preclude potato and vegetable cultivation. This constraint has been overcome by some households through the installation of jointly owned tubewells. Others have used the IRDP to obtain subsidized credit for a tubewell. Still others purchase water·from a neighbor's tubewell, often paying a higher than market rate for it by working as field-hands.

Case Studies

Three households from the marginal/near-landless category have been selected to illustrate strategies that involve cultivating high-value, labor-intensive crops.

In Jamalpur village marginal households constitute half the total. Figure 7.10 illustrates the income profile of the average marginal household and places the case-study households in context with the village average. The average size of the marginal landholder household is 5.3 persons. Household and per capita incomes are Rs 13,400 and Rs 2,500, respectively. Crop income constitutes more than one-third of total income, and dairying accounts for almost one-fourth. Agricultural labor and off-farm income also play a significant role, contributing nearly one-fifth each to the total.

Kailash Govind Household, Jamalpur. Kailash Govind, sixty years old and illiterate, heads an extended household of ten members. Of the six adult members in the work force, two are male and four are female. Kailash Govind and the four women engage in crop cultivation, manage the dairy enterprise, and hire out as agricultural laborers. The son commutes to Hapur, where he works six days a week as a mechanic in a sewing machine factory.

Figure 7.10 indicates that the annual household income, at nearly Rs 18,000, is higher than the average for the marginal category, whereas the per capita figure (Rs 1,800) is somewhat below average. This is because Kailash Govind's household is almost twice as large as the average marginal household. The son's earnings from the factory account for more than half the total household income.

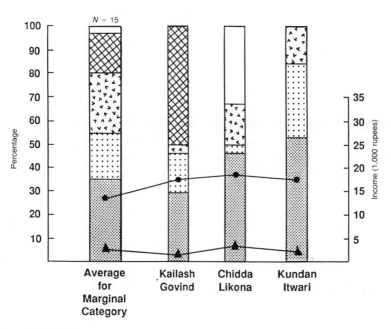

Source: Table A7.11.

Figure 7.10. Amount and sources of income for marginal/near-landless case-study households in Jamalpur, 1988–1989

Kailash's father was a medium landholder who owned about 2.8 hectares of land. His land was divided among his six sons, and Kailash Govind inherited 0.46 hectares. Until 1986, Kailash purchased water from his neighbor's tubewell. However, although the owner willingly sold water for three to four irrigations for the wheat crop, Kailash found it difficult to purchase water for potato or vegetable cultivation during the summer months. His brothers experienced the same problem. They could not diversify their crop operation without assured irrigation. In 1986 the six brothers installed a tubewell in partnership. This made possible the replacement of some wheat and maize with potatoes and vegetables.

Another factor, in addition to assured irrigation, that encouraged a reduction in the acreage planted to wheat and maize is the fact that Kailash Govind's son has a regular source of off-farm income. This permits them to purchase about 30 percent of their food requirements from the market. The remainder comes from their own production or in-kind payment for agricultural labor.

Kailash Govind's house reflects the improvement in the household's income. The house, consisting of three rooms and a courtyard, is of brick and cement. There is a hand pump in the courtyard and a domestic power connection. Livestock assets consist of one male and one female buffalo and a calf. The male buffalo is used for plowing. In 1988 the brothers invested in an electric thresher based on a one-sixth partnership for each. The thresher stands in one corner of Kailash's courtyard. In addition there is a manual fodder chopper. Consumer durables include a bicycle (used by the son to commute), a radio, a table fan, and a sewing machine. The household has accumulated an assortment of furniture, including a table and a couple of chairs.

Although he is a marginal landholder with less than 0.5 hectare of land, Kailash Govind has diversified the economy of his household. The tubewell partnership works because the brothers' six plots of land are close together and all within the command area of a small 5-horsepower tubewell. With his increasing confidence in potato and vegetable cultivation, Kailash Govind expects to increase the acreage under these crops.

Chidda Likona Household, Jamalpur. Chidda Likona, sixty-two years old and illiterate, heads a household of six. His oldest son, who has a high school diploma, is a teacher in a primary school in a neighboring state. The second son is a motor mechanic who lives and works in Meerut while his wife and three children stay with Chidda Likona in the village. He remits a portion of his earnings for the maintenance of his family.

Figure 7.10 shows that the annual income of the Chidda Likona household is more than Rs 19,300, and the per capita figure is above Rs 3,200. Both are significantly higher than the average for households in the marginal category. Over 45 percent of total household income comes from crop cultivation, remittances make up about one-third, and dairying contributes 16 percent. Because the women work on their own land and tend to the dairy enterprise, agricultural labor does not play a significant role in generating income.

Chidda Likona inherited 0.4 hectare of land irrigated by a traditional well. In 1960 he installed a Persian wheel. By the early 1980s, however, the water table had fallen so that irrigation with a Persian wheel became difficult. In 1982 Chidda borrowed from a private moneylender (at an interest rate of 30 percent per annum) and purchased a tubewell. A year later his son obtained work as a mechanic and was able to send home part of his earnings. The combination of assured irrigation and a reliable source of monthly income permitted a dramatic change in Chidda's farming operations. He had been cultivating wheat in the *rabi* season and maize and fodder in *kharif*, but the assured monthly remittances allowed Chidda to replace wheat in *rabi* with onions. In *kharif* he plants fodder for the dairy enterprise. He began purchasing all of his wheat from the market. Cultivation of onions, largely with family labor, brings a return per hectare more than twice that of wheat: Rs 11,800 rather than Rs 5,500. He sharecrops onions on an additional 0.35 hectare in the neighboring village. The landlord is a schoolteacher who is unable to devote much time to cultivation. The arrangement calls for the teacher to bear the entire cost of irrigation and chemical fertilizer while Chidda provides the seedlings, farmyard manure, and labor. They share equally in the output. By this arrangement Chidda Likona earns an additional Rs 3,500.

His house is made of brick. There is a hand pump in the courtyard. Although he does not have a domestic power connection, he confessed that he draws electricity illegally from the transmission lines. Consumer durables include a bicycle and a radio. Livestock assets include two buffaloes and two calves. The milk is sold to the local milk vendor. A strategy of educating his sons, diversifying into off-farm activities, and cultivating high-value crops has enabled Chidda Likona to improve his living standard despite a lack of land and capital resources.

Kundan Itwari Household, Jamalpur. Kundan Itwari, forty, has less than a primary education and heads an extended household of nine members: four adult men, two women, and three children. Although Kundan Itwari is a marginal landholder and owns less than 0.5 hectare, his reputation as a good farmer allows him to obtain almost twice this amount of land on a sharecropping basis. Family labor constitutes a crucial input to this extended operation. The household does not participate in any off-farm activity. In fact, Kundan Itwari once took out a subsidized loan under the IRDP to start a soap-manufacturing unit in the village. The venture was not successful, and Kundan and his family reverted back to crop cultivation. The loan has since been repaid.

Figure 7.10 shows that the household's annual income is above Rs 17,200, and the per capita figure is above Rs 1,900. The large family accounts for the latter being relatively low. Crop cultivation accounts for more than half of the household's income, agricultural labor for about a third, and dairying for the remainder. There is no off-farm income.

Kundan Itwari's land resources are very meager for so large a family, hence the household's strategy is to maximize returns to its labor resources. This is done through the cultivation of potatoes on leased land, hiring out as agricultural labor, and managing a dairy enterprise.

On his own 0.42 hectare of land Kundan plants wheat in the *rabi* season and fodder and maize in *zaid* and *kharif*. He does not have his own means of irrigation and therefore purchases water from his neighbor's tubewell. The neighbor, however, has difficulty in irrigating his own fields during the summer months because of frequent breakdowns in the supply of electricity and does not have much surplus water to sell. For this reason Kundan plants maize and fodder in *zaid* and *kharif* rather than paddy or potatoes. The wheat and maize from his land contribute a substantial proportion of the household's food supply; the remainder is obtained as payment in kind during the wheat harvest.

For the past three years Kundan Itwari has sharecropped potatoes on about 1 hectare of land in the neighboring village. The landlord owns about 4 hectares, on which he plants sugarcane, wheat, fodder, and potatoes. Because the landlord owns a tubewell, irrigation is not a problem. The landlord provides the irrigation, pays for preparation of the field, and shares equally in the cost of fertilizer. Kundan Itwari provides the seed and all the labor input. His income from potato cultivation on leased land is more than four times his income from wheat, maize, and fodder cultivation on his own land.

Although he belongs to the cooperative society, Kundan has been unable to get a loan to purchase potato seed to plant on the leased land. His sharecropping is by verbal agreement, and he therefore has no document to prove his lease. He has to borrow from a private moneylender at 30 percent interest to purchase his seed.

Kundan's house is part mud and part brick. There is a hand pump in the courtyard and a domestic power connection. Consumer durables include three bicycles, a radio, a table fan, a sewing machine, and a couple of wristwatches. Livestock assets include two buffaloes and two calves.

Kundan Itwari is a marginal landholder without his own means of irrigation. However, he has been able to overcome both the land and water constraints by sharecropping potatoes. Cultivation of this high-value, labor-intensive crop has led to increased use of family labor—the household's major resource.

Appendix

Table A7.1. Percentage distribution of landholdings and land owned in Jamalpur, by landholding group, 1976–1977 and 1980–1981

Landholding group	1976–77		1980–81	
	Number	Area	Number	Area
Marginal (<1 ha)	55	15	61	16
Small (1–2 ha)	21	19	22	22
Medium (2–5 ha)	17	36	9	27
Big (>5 ha)	7	30	8	35
TOTAL	100	100	100	100

Source: Jamalpur village land records, 1976–77, 1980–81, District Land Records Office, Collectorate, Meerut.

Table A7.2. Composition of Jamalpur work force, 1971 and 1981 (number of workers)

Workers	1971	1981
Agricultural	128	115
	97.0%	76.2%
Nonagricultural	4	36
	3.0%	23.8%
TOTAL	132	151
	100%	100%

Source: India, Office of the Registrar General, *Census of India*, General Population Tables, Series 1 (Delhi, various years).

Table A7.3. Gross cropped area under major crops in Jamalpur, 1951–1952 through 1988–1989 (ha)

Crop	1951–52	1979–80	1988–89
Wheat	41.3	89.0	90.7
Paddy	10.9	19.0	4.5
Maize	4.5	36.8	53.4
Pulses	53.4	7.3	23.0
Oilseeds	—	0.4	0.7
Sugarcane	34.8	65.6	31.0
Fodder	15.0	30.4	62.0
Vegetables	6.1	6.5	34.9
Fruits	1.0	6.9	23.9
Potato	2.0	14.1	52.0
Others	64.5	15.0	24.0
TOTAL	233.5	291.0	400.1

Source: Jamalpur village land records, 1951–52, 1979–80, 1988–89, District Land Records Office, Collectorate, Meerut.

Table A7.4. Net cropped, net irrigated, and tubewell irrigated area, in Jamalpur, 1970–1971, 1980–1981, and 1988–1989 (ha)

Area	1970–71	1980–81	1988–89
Net cropped area	192	193	195
Net irrigated area	192	193	195
Area irrigated by tubewells	100	154	195

Source: Jamalpur village land records, 1970–71, 1980–81, 1988–89, District Land Records Office, Collectorate, Meerut.

Table A7.5. Cropping intensity in Jamalpur, 1970–1971, 1980–1981, and 1988–1989

Area	1970–71	1980–81	1988–89
Net cropped area (ha)	193.0	192.6	194.8
Gross cropped area (ha)	355.0	380.8	450.5
Cropping intensity (percentage)	184	198	231

Source: Jamalpur village land records, 1970–71, 1980–81, 1988–89, District Land Records Office, Collectorate, Meerut.

Table A7.6. Man-days of hired agricultural labor per household, by village, 1988–1989

Village	Total agricultural labor income (Rs)	No. of landless and marginal households	Average wage rate of agricultural labor (Rs per man-day)	Man-days of agricultural labor per household
Walidpur	83,156	54	20	77
Rampur	34,432	20	20	86
Izarpur	22,791	26	20	44
Jamalpur	55,293	21	18	146

Sources: Tables A3.9, A3.10, A5.6, A6.11, and A7.9.

Table A7.7. Net and gross cropped area per household in Jamalpur, by landholding group, 1988–1989 (ha)

Landholding group	Net cropped area	Gross cropped area
Marginal	0.39	0.80
Small	1.94	3.76
Medium	3.62	6.89
Big	6.30	10.76

Source: Jamalpur household sample survey, 1988–89.

Table A7.8. Percentage distribution of gross cropped area in Jamalpur, by major crops and landholding group, 1988–1989

Crop	Marginal	Small	Medium	Big
Wheat	23.7	22.9	12.2	18.8
Paddy	—	—	—	—
Maize	18.8	16.8	6.1	7.8
Pulses	—	—	—	—
Oilseeds	—	1.9	3.0	17.0
Sugarcane	6.3	16.8	13.4	16.0
Fodder	30.7	21.6	—	25.0
Fruit	—	1.5	—	7.0
Vegetables	12.5	6.2	32.9	—
Potato	8.0	12.3	32.4	8.4
TOTAL	100	100	100	100

Source: Jamalpur household sample survey, 1988–89.

Table A7.9. Annual household income, per capita income, and percentage distribution of income in Jamalpur, by source and landholding group, 1988–1989 (rupees)

Landholding group	No. of households	Crop income	Noncrop income				Total household income	Per capita income
			Agricultural labor	Dairying	Off-farm employment	Rents, remittances, etc.		
Landless	6	333 (3.0)	2,733 (24.3)	1,533 (13.6)	6,392 (56.7)	271 (2.4)	11,262 (100)	2,048
Marginal	15	4,783 (35.6)	2,593 (19.3)	3,349 (24.9)	2,319 (17.2)	420 (3.1)	13,464 (100)	2,525
Small	6	20,866 (67.9)	—	5,993 (19.5)	1,100 (3.6)	2,750 (9.0)	30,708 (100)	2,835
Medium	2	33,391 (78.4)	—	2,220 (5.2)	3,600 (8.4)	3,400 (7.9)	42,611 (100)	5,013
Big	1	66,746 (94.4)	—	3,950 (5.6)	—	—	70,696 (100)	7,070
TOTAL	30	11,082 (54.5)	1,843 (9.1)	3,459 (17.0)	2,898 (14.3)	1,041 (5.1)	20,323 (100)	2,974

Source: Jamalpur household sample survey, 1988–89.
Note: Figures in parentheses are percentages of total household income.

Table A7.10. Cumulative percentage of aggregate household income in Jamalpur, by source, 1988–1989 (N = 30)

Landholding group	Households	Crop income	Noncrop income	Total income
Landless	20.0	0.6	23.7	11.1
Marginal	70.0	22.2	70.7	44.2
Small	90.0	59.9	92.0	74.4
Medium	96.7	80.0	98.6	88.4
Big	100	100	100	100

Source: Jamalpur household sample survey, 1988–89.

Table A7.11. Annual household income, per capita income, and percentage distribution of income, by source, for marginal case-study households in Jamalpur, 1988–1989

Income per household	Average of marginal category		Kailash Govind household		Chidda Likona household		Kundan Itwari household	
	Rupees	Percentage	Rupees	Percentage	Rupees	Percentage	Rupees	Percentage
Crop income (own land)	4,093	30.4	5,123	28.8	4,751	24.6	1,625	9.6
Crop income (leased land)	687	5.1	—	0	4,150	21.5	7,500	43.4
Agricultural labor	2,599	19.3	3,000	16.9	800	4.1	5,400	31.3
Dairying	3,353	24.9	607	3.4	3,130	16.2	2,720	15.8
Off-farm employment	2,316	17.2	9,060	50.9	—	—	—	—
Rents, remittances, pensions, interest	417	3.1	—	—	6,500	33.6	—	—
Total income per household	13,464	100	17,790	100	19,331	100	17,245	100
Income per capita	2,525		1,779		3,222		1,919	
Main occupation			Off-farm employment		Crop cultivation		Crop cultivation	
Subsidiary occupation			Crop cultivation		Remittances		Agricultural labor	

Sources: Jamalpur household sample survey and Jamalpur case-study household survey, 1988–89.

8

Speeding up Income and
Employment Diffusion:
Some Policy Implications

The purpose of the study on which this book is based was to reappraise the Green Revolution in the light of early assessments and subsequent developments. The early assessments, while hailing its favorable impact on output, lamented the considerations of equity it raised: an apparent widening of interregional and interpersonal disparities and a further marginalization of the rural poor. The findings of our study challenge this assessment and suggest that while the initial impact of technical change indeed supported a pessimistic point of view, subsequent developments do not. Through an examination of changes that have occurred in the state of Uttar Pradesh, especially the western region, one of the principal locales of the Green Revolution, we have demonstrated that the interregional and interpersonal impacts of the new technology have varied over time. Our findings indicate, first, that while regional disparities widened in the initial phase of the post–Green Revolution period—from the mid-1960s to the mid-1970s—subsequent years witnessed a narrowing of differences as technology diffused into the hitherto bypassed regions in the wake of infrastructure development. Second, the second-generation effects of the Green Revolution began to be reflected in a variety of noncrop and off-farm employment opportunities and in the growth of small and medium rural towns that increasingly served as the focus of rural off-farm activity. The growing participation of the landless and near-landless households in these activities has led to an improvement

239

in their income levels and dispelled the notion that technological change pauperizes the rural poor.

Summary of Findings

We examined the impact of the Green Revolution technology at three levels: the region, the village, and the individual household.

Regional-Level Comparison

Macro data were used to examine the initial widening and subsequent narrowing of regional disparities between western and eastern U.P. Our study shows that the impact of HYV technology was first felt in western U.P., where the foundation for commercial agriculture had been laid more than a century ago with the opening up of the Upper Ganga and the Eastern Yamuna canal systems. The modern varieties introduced in the mid-1960s not only yielded impressive quantities of wheat, they transformed the nature of the irrigation system as well. Canal irrigation, the mainstay of the agriculture of the Upper Gangetic plain for more than a hundred years, gave way to private tubewells as the major source of irrigation. Tubewells gave cultivators the ability to control the application of water, essential to the HYV package of practices and a prerequisite to the multiple cropping that followed.

Eastern U.P., despite its fertile soil and adequate water resources, was initially not as receptive to modern technology. Two factors—a deficiency in irrigation and the more feudal agrarian structure inherited at independence—were responsible. Both were soon corrected, and by the mid-1970s, a decade after the west, agricultural growth in eastern U.P. had picked up momentum. Growth of agricultural output for major crops in eastern U.P., less than 2 percent per annum between 1962–65 and 1970–73, rose to more than 4 percent per annum for the years between 1970–73 and 1980–83, with six of the fifteen districts recording rates higher than 5 percent. Agricultural growth in the west, on the other hand, began to slow during the second period, dropping from 4.5 percent per annum to 3.8 percent.

It is clear that the development process derives a large share of its stimulus from agricultural growth. Western U.P., which experienced the early impact of the Green Revolution, also shows clear evidence of a diversifying rural economy. In the east, in contrast, the rural economy is still dominated by crop cultivation, and diversification into off-farm activities is not as visible.

Village-Level Studies

Micro-level research was carried out in four villages of Meerut District in western U.P. In Walidpur, for which a comparison over time was possible, we found that both total income and its distribution had improved. Three mechanisms by which the poorer households benefited were identified: off-farm employment, dairying, and labor-intensive cultivation of high-value crops. We then examined the operation of each of these mechanisms in a case-study village. Unfortunately we do not have time-series data for these three villages so it is not possible to quantify how much income has increased over time. But the evidence from Walidpur points to a decline in the incidence of poverty and suggests that the real income of households in the marginal and small landholding categories may have increased by 80 and 73 percent, respectively, between 1963–64 and 1988–89. The diversifying income structure, with a rising share of off-farm earnings for landless and near-landless households—wage rates in off-farm employment being higher than in agricultural labor—implies that the absolute income of the rural poor has improved in the other villages as well.

Household Case Studies

The study was carried one step further in order to understand how individual households respond to the changing opportunities brought about by the new agricultural technology. Our case studies among landless and near-landless households illustrate some of the strategies they have devised to overcome constraints caused by lack of land and capital. The strategies adopted by these households suggest areas of possible policy intervention by which the income and employment diffusion process could be hastened, including the following:

- Strengthening roads and transportation
- Improving rural electrification
- Fostering growth centers
- Streamlining credit services
- Developing technical skills in rural youth

These are discussed in detail later in this chapter.

Is the Case of Meerut and Western Uttar Pradesh Applicable to Other Regions of India?

The creation of new employment opportunities for the rural poor observed in our study is not unique to Meerut and western U.P. The Punjab, the heartland of the Indian Green Revolution, has experienced a similar economic diversification. G. K. Chadha (1983:172) noted that

> the Punjab has perhaps the most spatially balanced urban structure among Indian states, for the towns are distributed among the state fairly evenly. Each town in Punjab is a grain market and is linked with its rural hinterland. The addition of modern factory industries is only an extension of such rural urban links. . . . A variety of industries are located in the villages and semi-urban places in response to local demand and entrepreneurial expertise. Typical village industries are engaged in rope, carpet, *gur* and *khandsari*, soap, shoe and pottery making, handloom weaving, oil-crushing, tanning, processing of cereals and pulses, and so on. . . . In recent years, rice shelling has emerged as a very important industrial activity in a number of villages, especially the larger ones, and in some central villages, which cater to the requirements of a cluster of villages around, repair shops are very common.

The similarity in the nature and growth of off-farm activity in Punjab, western U.P., and Meerut—regions that experienced the first-round effects of the Green Revolution—suggests that diversification has been stimulated by the second-generation effects of the new agricultural technology. An alternative explanation, that the growth of the noncrop and off-farm economy in the region derives stimulus from its proximity to Delhi, does not appear to be valid, for there is no evidence of similar patterns of rural diversification in regions around other major cities such as Calcutta and Bombay. The immediate conclusions of this study are therefore applicable to areas where agroclimatic conditions and other socioeconomic factors have favored the adoption of the Green Revolution technology, principally the western Indo-Gangetic plain and the coastal areas of peninsular India.

The experience of Meerut and western U.P., however, cannot be immediately replicated in the regions of the middle and lower Gangetic plain. While these regions have immense potential due to their rich alluvial soils and abundant surface and subsurface water resources, they also suffer from special problems of waterlogging, frequent flood damage, and fragmented landholdings. The experience from western

U.P. will be replicable here only after suitable institutional changes have been effected. Strategies that incorporate water management, consolidation of landholdings, and development of rural infrastructure will take time to implement.

As for the dryland rainfed tracts of the Deccan, where water-fertilizer-seed technology has not made much of an impact, these regions will either have to await technological breakthroughs in dryland farming or make large investments in irrigation that will allow them to adopt the HYV wheat and rice technology.

Hastening the Income Diffusion Process

The salient feature of the second-generation effects of technological change is the expansion of off-farm employment opportunities and the increasing participation of landless and near-landless households in them. In all the study villages, wages from off-farm employment for unskilled and semiskilled workers range from one and a half times to three times wages derived from agricultural labor. In Walidpur village, for which the composition of household income can be compared over time for the marginal category, off-farm earnings increased from 28 percent of the total in 1963–64 to 44 percent in 1988–89. In all four villages, off-farm earnings contribute about 60 percent of total income in the landless class of households and about 45 percent in the case of the marginal category. The average per capita income of the sampled landless households in the four villages ranges from 20 percent to 50 percent above the official poverty line. The Walidpur data suggest that there has been a decline in the incidence of poverty between 1963–64 and 1988–89. The proportion of households with incomes below the poverty line in Walidpur, Rampur, and Izarpur is about 10 percent, and around 13 percent in Jamalpur.

The common characteristic of off-farm income observed in all the study villages is its tendency to raise the income levels of the poorer households and thereby reduce inequality between the various landholding groups. While crop income varies directly with the amount of land owned, and therefore tends to benefit the big landholders, off-farm income exhibits an inverse relationship to farm size and favors households at the lower end of the income spectrum.

In addition to raising the household income level of landless and near-landless households, off-farm income also has a salutary effect on income distribution. Decomposition of the Gini coefficient for income distribution in each of the four villages showed that the elasticity of inequality for off-farm income had the highest negative value, implying that marginal increases in earnings from this source would lead to the greatest decline in income inequality between various landholding categories. If the process of income diffusion is to be speeded up, it would be well to look at possibilities of increasing off-farm employment.

Encouraging the Rural Off-Farm Sector

In addition to its potential for reducing income inequality between rural groups, the nature of off-farm activities is such that they offer the prospect of diverting labor away from agriculture while simultaneously reducing the need for rural–urban migration. Although the majority of households in all the study villages are involved in off-farm activities, their area of operation is limited to within commuting distance of the village; migration to big cities or industrial towns, especially among the landless and near-landless households, is virtually nil.

Despite the potential of rural off-farm activities for reducing underemployment, this subject has not merited much attention from policymakers. Although rural off-farm activities are very visible to the eye in large villages and rural market towns, they are less apparent in official statistics. The very limited documentary evidence on the extent and diversity of the off-farm sector reflects the fact that the phenomenon has picked up momentum only over the last two decades. Moreover, many of the microenterprises—pushcarts, street vendors, roadside kiosks, tea stalls, *tempo-taxis*, agroservice centers, small wayside garages, and workshops—operate in the informal sector and are not registered anywhere. The activities, by their very nature, are widely dispersed, which makes them difficult to identify. Frequently they are carried on within the village with no clear-cut division between time devoted to off-farm and farm activities.

To the official eye, rural industry, especially traditional cottage industry (handloom and handicrafts), tends to be synonymous with the rural off-farm sector. However, off-farm activities in western U.P. are much wider in range and include household industry, agroprocessing, small manufacturing, trade, transport, and construction. Policies that target only rural industry thus relate to only a fraction of the rural off-farm sector.

Documentation for the Seventh Plan (India, Planning Commission 1985) contains sections dealing with the agricultural sector, rural industry, and rural development. In the latter section antipoverty programs that target households below the poverty line are emphasized. Our study indicates that the class of small rural entrepreneurs emerging from the landless, marginal, and small landholder categories do not necessarily fall below the poverty line and are able to invest their small surpluses, aided by informal credit channels, into off-farm enterprises. While aspects of the newly emerging off-farm sector are captured piecemeal under various different headings of the plan document, the existence of this sector as a major source of off-farm employment is ignored. The Eighth Five-Year Plan is scheduled to commence in 1992–93. The approach paper to the Eighth Plan targets an average GDP growth rate of 5.6 percent and declares employment generation to be one of its major objectives. Detailed strategies for employment generation are not spelled out, and the details of the plan document were in the process of being finalized when our study was completed. Hence it is not possible to assess whether the informal rural off-farm sector will receive greater attention during the Eighth Plan period than it has in the past.

The first step in the process of strengthening the off-farm sector is to recognize its existence and growing importance in the rural economy. Thanks to the work of Hernando de Soto (1988), the contribution of the "informal sector" in developing countries the world over is coming to be appreciated. In Peru the informal sector employs an estimated 48 percent of the total labor force, contributes 38 percent of GDP, operates 93 percent of the buses, and accounts for 42 percent of the housing. In Bolivia, Ecuador, and Nicaragua it employs about half of the working population (*Economist* 1988).

The emerging rural off-farm sector in Meerut and western U.P. is stimulated by a growth process different from the one in Latin America. However, several common problems could be mitigated through government policies that diminish official harassment and discrimination against the small rural entrepreneurs and instead facilitate their access to credit, services, inputs, markets, and skills.

Importance of Roads and Transportation. The opening up of the rural hinterland through a network of roads, transport, and communications has stimulated the expansion of both farm and off-farm activities.

The spread of vegetable and potato cultivation in villages off the main highways has been facilitated by the network of roads and transport. Improved rural roads linking villages to the highways have also had a positive influence on the spread of dairying. In Izarpur alone there are four private milk traders and a cooperative milk collection center. The private traders market most of their milk in Meerut City, 23 kilometers away, transporting it by bicycle, motorcycle, and horse cart. The milk from the cooperative center is collected twice a day in small trucks.

Both public and private transport have improved the mobility of the village population. Examples from our case-study households show that a large number of households supplement their income through labor in construction activities. Workers in this field usually commute to construction sites in private trucks and *tempo-taxis*. Often labor contractors come to the villages with trucks to transport workers to the construction site.

Moreover, several of the case-study households have boosted their incomes through the growing transport sector. Some members have found wage employment as truck drivers and bus conductors. Self-employment through investment in cycle-rickshaws, horse carts, and *tempo-taxis* for hire has added to the income of others. The transport sector is sustained by a large network of service establishments in which many members of our case-study households find employment.

While much has been accomplished, there are still many villages poorly connected by roads or inadequately serviced by transport. The local Panchayati Raj Institutions could play a more constructive role in mobilizing village labor and capital to correct this.

Improving Rural Electrification. Improved electricity service is essential for tubewell irrigation, but it is equally important for the small agroprocessing units that are being set up throughout the rural areas. Our case-study households indicate that increased incomes have led to a demand for custom milling of foodgrains to replace the drudgery of manual grinding. All the study villages have at least one power-driven *atta-chakki*, and one village has three. Villages that grow sugarcane also have small power-driven *kolhus* for crushing cane. In areas where short-duration oilseeds are making a comeback, small power-driven oil pressers are being installed. In addition to such processing units, small-scale industrial enterprises are also increasingly common within small villages. Such enterprises include cloth weaving by powerloom and the manufacture of small machinery parts using power-driven lathes.

Despite the fact that rural electrification in western U.P. played a crucial role in the Green Revolution of the 1960s and 1970s through the spread of tubewells, it is now a major factor inhibiting diversification of the rural economy. Although generation of power in the state increased almost twenty-five-fold between 1960 and 1990, the gap between demand and supply continues to widen. Rural areas in Uttar Pradesh receive power only eight to ten hours a day, and this is frequently accompanied by breaks in transmission and fluctuations in voltage. Severe voltage surges often result in burned-out transformer equipment and damage to tubewell motors.

In several instances among our case-study households the power supply situation has deterred entrepreneurs from starting a rural industry, caused losses in income, or constrained expansion of the enterprise. For instance, the Banwari Chinta household in Sitapur, whose major source of income is from the custom milling of wheat on a power-driven *atta-chakki*, suffered a loss equivalent to one-fifth of its annual income when a voltage surge burned out a transformer which was then left unrepaired for more than three months. Subedar Singh in Rampur, whose sewing thread business is a model of household entrepreneurship, was reluctant to expand the unit because of the uncertain power supply. Kathu Jugnu in Izarpur village wanted to invest in a power-driven oil mill but was dissuaded from doing so by the erratic power supply.

These are not isolated incidents; similar complaints were heard in many villages. In fact, the Bhartiya Kisan Union, an influential farmers' organization in western U.P., staged demonstrations on several occasions during 1987–89 to demand improvement in power availability and maintenance. In January 1988 an estimated 500,000 farmers gathered in Meerut to protest, among other things, the high electricity tariff, which they felt was not justified by the quality of service. While most of the protesters were farmers, there were also rural powerloom weavers in the group (Gupta 1988).

Alternate sources of energy, such as bio-gas and solar energy, have not made any significant impact in meeting rural energy requirements. Bio-gas plants are in operation in several villages, but the gas is used primarily for lighting and cooking purposes. Problems of design and nonfermentation of the biomass during winter have prevented the widespread application of this technology. Solar energy is still too expensive to compete with the conventional sources.

Fostering Rural Growth Centers. Rural off-farm employment offers an alternative to the classical European development process in which surplus agricultural labor was absorbed in large urban-based industrial complexes. It also differs from the experience of Japan and Taiwan, where export-oriented manufacturing units have played a key role in off-farm diversification. The experience of Meerut indicates that most rural off-farm activities at the present stage of development are expanding primarily in response to local demands for goods and services, and agriculture continues to be the main force behind this diversification.

Rural growth centers, usually small and medium-sized towns with populations ranging from 5,000 to 50,000, have aided in the process of diversification by providing a nucleus for off-farm activities. The oldest of these growth centers developed around rural market towns. In the 1930s and 1940s the establishment of private sugar factories created additional centers. The establishment of block development offices and primary health centers by the government in the 1950s added to their number and further strengthened the rural infrastructure. More recently, in the 1970s and 1980s, industrial estates and market yards developed by the government to encourage small industry and provide marketing facilities have evolved into growth centers.

Our case-study households exemplify cases in which the nearby growth center has provided employment for construction labor, unskilled and semiskilled workers in agroprocessing units, small manufacturing units, agroservice centers, and tailor and barber shops. Many young men in the village commute regularly to the growth center to attend high school or intermediate-level college. Nafeez Usman of Sitapur supplements his income by operating a cycle-rickshaw between his village and the nearby growth center, not only ferrying passengers but also regularly carrying village children to school in the town. Dilawar Singh of Rampur has invested in a *tempo-taxi* to carry both agricultural produce and passengers between Rampur and the market town of Dhaulri about 10 kilometers away. Horse carts and tractor-trolleys plying between villages and nearby growth centers are quite common.

The development of small and medium-sized towns as centers of rural off-farm activities has not received much attention from official agencies. The Meerut Development Authority focuses its attention on Meerut City only and concentrates on housing construction (Meerut, Development Authority 1989). The rural off-farm sector would benefit greatly if the Meerut Development Authority could extend its coverage to all urban areas within the district. Planning for small and medium-sized towns in the district should focus on on rural development as well as urban development. Several of the large villages in the district are approaching a critical mass in terms of population and diversification of their economy. Such future centers of growth need to be identified, and steps must be taken to equip them with the infrastructure and essential services required of growth centers. This should include the development of designated sites for shops, garages, workshops, and agroservicing, as well as the provision of such facilities as bank branches and a more assured supply of electricity. Planning should also focus on strengthening communication linkages between such growth centers and the cluster of villages around them.

Streamlining Credit Services. Access to capital is one of the major constraints faced by households trying to become self-employed in off-farm occupations. The more enterprising of these households overcome this constraint in part through savings from steady off-farm wage employment of one or more family members and in part through borrowing from private sources. This borrowing often entails very high interest rates, ranging from 24 to 48 percent per annum.

Those who qualify for loans under the IRDP can obtain credit at an interest rate of 11 percent as well as a subsidy of one-third to one-half of the loan amount. Qualified borrowers are supposed to have incomes below the poverty line. But a large number of households in the landless, marginal, and small landholder categories have incomes above the poverty line and hence do not qualify for cheap credit. It is primarily from this category of households that the rural entrepreneur class is emerging. Our case studies, notably Dilawar Singh and Subedar Singh of Rampur and Samay Singh of Sitapur, indicate that households that have been successful in starting a rural industry or business usually have incomes above the poverty line and have mobilized capital through a combination of household savings and private credit. These households have been unable to avail themselves of institutional credit. Potential borrower households have been discouraged by lengthy and complex bank procedures. Akhtar Ali of Rampur, who has the necessary technical skill and wants to set up a welding workshop, is unable to do so because of financial constraints.

There are other limitations to the institutional credit procedure. Bank loans to smallholders are customarily tightly supervised because the borrower is presumed to have minimal management capabilities. In order to ensure that funds are not used for consumption purposes, bank checks are frequently made out to specific dealers rather than to the beneficiary. This can have the effect of dampening the borrower's enthusiasm to manage the enterprise for himself and shop around for the cheapest supplier. Lending procedures also call for two land-owners from the village to stand as guarantors for the borrower. Bor-rowers are reluctant to incur such obligations as it often places them in dependency relationships with the landowners. Loans in excess of Rs 5,000 require collateral, and this is a major obstacle for landless and near-landless households.

There is a strong case to be made for simplification of lending pro-cedures, and especially for review of procedures that contribute to frustration, delay, and malpractice. It is questionable whether these elaborate precautionary measures result in any commensurate gain in recovery rates.

Another possible route to credit is through group borrowing. The theory is that group borrowing improves the loan recovery rate through joint responsibility and reduces the borrowers' and lenders' transaction costs. This has been the experience of the highly success-ful Grameen Bank in Bangladesh. However, there has been little expe-rience with group borrowing in Meerut. On the other hand, coopera-tion among marginal landholders to purchase assets is gaining ground as a strategy to overcome the capital constraint. Of our case-study households, Kailash Govind of Jamalpur was able to start cultivating high-value crops after he purchased a tubewell jointly with five other marginal farmers; Samay Singh of Sitapur also jointly owns a tube-well. Bashir Mohammad of Sitapur established a grain mill in the village in partnership with two other households. This unit required an initial investment of more than Rs 10,000, an amount almost equal to the annual household income of an average landless house-hold. Joint ownership of assets has increased considerably in the past decade or so. Banking institutions should devise systems to encourage this new tradition.

Developing Technical Skills. Although formal education is reduc-ing illiteracy throughout rural India, access to employment oppor-tunities is more likely to increase through development of technical skills. The nature of the rural off-farm sector is such that those with some technical training are more likely to procure wage employment and start their own enterprises. Government-run industrial training institutes and polytechnics offer technical training for welders, fitters, garage mechanics, electricians, and carpenters. There are four such institutions in Meerut District with facilities for training about 2,500 persons per year. They require at least a high school–level education for admittance.

Several of our sample households improved their household income despite little or no formal education, largely through acquiring an intermediate level of skills through the traditional apprenticeship method. Hashim Ali and Dilawar Singh of Rampur, Bashir Mohammad and Sachdeva Mahato of Sitapur, and Kailash Govind of Jamalpur are cases in point. Many of the younger generation in the study villages are learning to drive trucks or are apprenticed to furniture makers, garage mechanics, welders, electricians, and tailors. This class of rural youth needs a system that can incorporate the traditional apprenticeship method into an informal institutional arrangement. The government-run TRYSEM program was started with this intention; however, funding restrictions have limited its scope. Until recently only about 700 young people benefited from the program each year.

Encouraging Animal Husbandry

Although we selected Izarpur village to demonstrate the positive effect of dairying on the level and distribution of household incomes, we found that dairying income played a significant role in supplementing the household income of the poorer households in the other three study villages as well. Unlike land, dairy animals are equally distributed among the various rural groups. Dairying income constitutes between 15 and 30 percent of the total in the landless and near-landless groups in the study villages. In addition to its contribution toward improving the income levels of the poorer households, income from dairying has the least unequal distribution.

Small producers have been able to participate in dairying on an appreciable scale only since about 1975. The second-generation effects of the Green Revolution have increased rural incomes, thereby allowing the landless and near-landless to invest some of their savings in the dairy enterprise. The experience of Meerut also indicates that with the almost universal use of hand pumps for water, together with mechanization of threshing and winnowing and milling operations of foodgrains, women have been released from much of the drudgery of traditional chores, and they can devote more time to dairying and thereby supplement the household's income.

Our case studies indicate that the household dairy enterprise is influenced by availability of female family labor, capital for the purchase of a dairy animal, and access to fodder. It is noteworthy that credit for the purchase of a dairy animal is relatively easier to obtain than credit for starting an off-farm enterprise. Our case-study households indicate that the local *dudhiya* frequently provides credit for the purchase of a buffalo, as well as for purchase of fodder and concentrates, provided the milk is sold to him at rates that are usually lower than the market rates. Milk cooperatives have not been very successful in eliminating the institution of the *dudhiya*; nor does it appear necessary to eliminate the *dudhiya* altogether. The spread of milk cooperatives can help to make the milk market more competitive. In Izarpur, where the cooperative milk center functions well, the *dudhiya* pays between 15 and 30 percent less for milk than the cooperative society. In villages without milk societies and with poorer access to markets owing to distance from market centers, the rates offered by the *dudhiyas* are often half of what cooperative societies would pay.

The main constraint faced by landless and near-landless households in increasing their dairy operations is limited access to fodder. Presently the bulk of their fodder requirement is met through in-kind earnings of sugarcane tops as wages for harvesting sugarcane, straw obtained as payments during the wheat harvest, and by weeding the fields of medium and big farmers in exchange for the weeds and grasses that can be carried away.

Research and extension input for dairying is available to a much lesser extent than for crop cultivation. The low priority accorded to dairying and animal husbandry can be assessed from the allocation of funds to this sector. The First State Plan (1955–56) allocated less than 1 percent of total funds to animal husbandry. For the Sixth Plan (1980–85) the share of this sector declined to about 0.5 percent (Uttar Pradesh, State Planning Institute, *Sankhyiki Patrika, Meerut*).

There is a need to improve productivity of buffaloes through better breeding and better feeding, health care, and management. The average buffalo in the village yields about 5 liters of milk per day, while improved breeds yield about twice that amount. Although there are over a hundred artificial insemination centers in Meerut District, small producers rarely use them, mainly because of the low rate of conception—often less than 25 percent. This means that an owner must take his buffalo to the center at least three or four times. A more effective system would be coverage of all female buffaloes through mobile teams. This, however, will not be possible until the centers are provided with the necessary transport. While improvement of breeds is a long-term process, milk yields can be improved in the short run through improved feeding.

Given the demonstrated potential of dairying for raising income levels of landless and near-landless households, improving village income distribution, and providing gainful employment for female family labor, there is a clear need to accord the dairying sector a much higher priority in allocation of state funds than it has received thus far.

Improving Government Antipoverty Programs

Of the various antipoverty programs operating in Meerut District, the Integrated Rural Development Program has been the most influential in the four study villages. Although there was no way for us to compare income levels of beneficiary households before and after acquisition of IRDP loans, it is clear that earnings from buffaloes purchased under the program, or from off-farm activities such as a *parchoon* shop or a buffalo cart, increased income levels of the landless and near-landless households.

In Izarpur, the village chosen for its dairy enterprises, about 40 percent of landless and marginal households have been beneficiaries under the program. Some households availed themselves of the maximum loan of Rs 9,000 with the maximum subsidy benefit of Rs 3,000; others took only half this sum. More than 70 percent of those who purchased a milk animal in the past five years still have either the original asset or its progeny. The other 30 percent had to sell their animal for cash to meet a household consumption requirement.

The perception of the program among borrowers is favorable, despite problems of procedural delay and transaction costs. The past repayment of loans in Izarpur shows an above-average rate of recovery. About 55 percent of the sample households who benefited under the scheme have repaid the loan. Of the remaining, about 30 percent are in various stages of repayment. About 15 percent, mostly recent beneficiaries, are in default, laboring under the impression that the loans will be written off in the near future, a notion propagated by some political parties preceding the 1989 general elections. Our case-study households highlight two drawbacks in the implementation of the IRDP program: the misidentification of potential borrowers and the lack of monitoring and follow-up.

Households with income levels below Rs 6,400 (the official poverty line for selection of beneficiaries in 1988–89 [Meerut, District Rural Development Agency 1989]) are supposed to be given priority in selection. Some of the case-study households that received loans under the program for a second and third time are visibly above poverty line status, while several households in the village that appear more deserving have been ignored. At present only the village *pradhan* (elected headman) is involved in the selection of beneficiaries, although guidelines call for a general village meeting to be held to identify the poorest among them. Some households who did not vote for the *pradhan* feel they were automatically eliminated from the list of potential beneficiaries. A less biased procedure for selection is clearly required. This could involve cross-checking by village consensus in an open meeting.

As our case-study households show, sometimes a cottage industry fails after the household has obtained a loan for it under the IRDP. Kundan Itwari of Jamalpur village tried to diversify into off-farm activities but was obliged to revert to crop cultivation after his enterprise failed. There was no follow-up by any agency to identify causes of failure.

There is no doubt about the potential of the IRDP to improve the income levels of the landless and near-landless; it is one of the few governmental programs actively encouraging the process of noncrop diversification. However, the program would be more effective in poverty alleviation if greater attention were paid to the selection of beneficiaries.

Summing Up

It appears, then, that the Green Revolution can be a double-barreled blessing for India. Not only does it hold the potential for increasing foodgrain production at rates ahead of population growth; it also bids fair to help resolve the country's far more perplexing employment problem.

The experience of Meerut and western U.P. suggests that the rural poor can benefit indirectly from technical change in agriculture. Their limited land base prevents landless and near-landless households from reaping the benefits of the seed-water-fertilizer technology to the same extent as the big landholders. However, their gains from the diversification of the rural economy can to a significant extent compensate for this. Some of this diversification will be into the noncrop farm sector. But it is the rural off-farm sector, increasingly centered in small and medium-sized towns, that affords the most attractive opportunities for employment and income diffusion, midway between the agricultural and the urban-industrial sectors.

Rural diversification's promise of relief for the employment dilemma should not, however, be taken as grounds for complacency. For the moment it applies only to those portions of India where rapid agricultural change is possible, and these still constitute but a fraction of the country. Further, there is the staggering dimension of the demographic problem that underlies the need for more jobs. India's population—at over 800 million more than double that counted at independence—continues to grow by nearly 17 million each year, and, at 32 per thousand of population, the birthrate continues to be unacceptably high. Only after the birthrate has been brought down can a lasting solution be truly at hand.

Appendix

Decomposition of the Gini Coefficient
by Sources of Income

Lerman and Yitzhaki (1985) showed that the Gini coefficient can be derived directly from the formula for Gini's mean difference

$$A = \int_a^b F(y)[1 - F(y)]dy \qquad (1)$$

where y represents income, a the lowest income, b the highest income, and F the cumulative distribution of income. Using integration by parts and variable transformations, they demonstrated that

$$A = 2cov[y, F(y)]. \qquad (2)$$

Then the Gini coefficient (G) for total income is obtained by dividing equation 2 by the mean income, m (Stuart 1954).

Letting $y_1 \ldots y_k$ be components of household income, such that

$$y = \sum_k y_k$$

and using the properties of the covariance,

$$A = 2 \sum_k cov(y_k, F(y)) \qquad (3)$$

where $cov(y_k, F(y))$ is the covariance of income component k with the cumulative distribution of income. Dividing equation 3 by m (obtain-

256

ing the relative Gini) and then multiplying and dividing each component k by $cov(y_k \, F(y))$ and by m_k yields the decomposition of the Gini coefficient by source.

$$G = \sum_k [m_k/m] \cdot [2 \, cov(y_k, F(y_k))/m_k] \cdot [cov(y_k, F(y))/cov(y_k, F(y_k))] \quad (4)$$

$$G = \sum_k S_k G_k R_k \quad (5)$$

where S_k is share of component k in total income, G_k is the relative Gini of component k, and R_k is the "Gini correlation" between income component k and total income.

Equation 5 enables us to decompose the role of a particular income source—for instance, off-farm employment—in inequality into three easily interpretable terms:

• The magnitude of off-farm earnings relative to total income
• The inequality of off-farm income
• The correlation of off-farm income with total income

Pyatt et al. (1980) showed that the Gini correlation (R) ranges between -1 and $+1$. Thus R will equal $1(-1)$ when an income source is an increasing (decreasing) function of total income. R will equal zero when the income source is constant, indicating that the source's share of inequality is zero.

An important criterion for studying decomposition by source is to understand how changes in certain income sources influence overall income inequality. To determine the elasticity of inequality, defined as change in inequality due to marginal change in a particular income source y_k, Lerman and Yitzhaki (1985) and Stark et al. (1986) considered a change in each household income from source k equal to $e_k \, y_k$, where e_k is close to 1. Then,

$$\frac{\partial G}{\partial e_k} = S_k(R_k G_k - G) \quad (6)$$

and

$$\frac{\partial G}{\partial e_k}/G = \frac{S_k R_k G_k}{G} - S_k \quad (7)$$

The elasticities in equation 7 sum to zero because a proportional increase in income from all sources would leave income inequality unaffected (Boisvert and Ranney 1990). Equations 6 and 7 show that so long as a particular income source constitutes part of the total income, for example off-farm earnings, then the following apply:

- If the Gini correlation between off-farm income and total income (R_k) is negative or zero, an increase in off-farm income *necessarily decreases* inequality
- If the Gini correlation is positive, then the impact on inequality depends upon the sign of ($R_k G_k - G$). A necessary condition for inequality to *decrease* is that inequality of off-farm income must be less than the inequality of total household income:

$$G_k < G \text{ since } (R_k \leq 1).$$

References

Adelman, Irma. 1984. "Beyond Export-led Growth." *World Development* 12: 937–949.

Adelman, Irma, and J. E. Taylor. 1990. *Changing Comparative Advantage in Food and Agriculture: Lessons from Mexico.* Paris: Development Centre of the OECD.

Agarwal, V. K. 1988. *Marketing of Dairy Products in Western U.P.* Bombay: Himalaya Publishing House.

Ahluwalia, Montek S. 1986. "Rural Poverty, Agricultural Production and Prices: A Re-examination." In J. W. Mellor and Gunvant M. Desai, eds., *Agricultural Change and Rural Poverty: Variations on a Theme by Dharm Narain.* New York: Oxford University Press.

Alagh, Yoginder L. 1982. "Indian Agricultural Economics—Some Tasks Ahead." Inaugural address delivered at the 42nd annual conference of the Indian Society of Agricultural Economics. *Indian Journal of Agricultural Economics* 37 (4): 421–425.

Bala, Raj. 1986. *Trends of Urbanization in India.* Jaipur: Rawat Publications.

Bandyopadhyay, D. 1988. "Direct Intervention Programmes for Poverty Alleviation: An Appraisal." *Economic and Political Weekly* 23 (26): A77–A88.

Bardhan, Pranab. 1970. "Green Revolution and Agricultural Labour." *Economic and Political Weekly* 5 (29–31): 1239–1246.

———. 1973. "Variations in Agricultural Wages: A Note." *Economic and Political Weekly* 8 (21): 947–950.

———. 1974. "Inequality of Farm Incomes: A Study of Four Districts." *Economic and Political Weekly* 9 (6–8): 301–307.

———. 1986. *Land, Labor and Rural Poverty: Essays in Development Economics.* New Delhi: Oxford University Press.

Barker, Randolph. 1987. "A Perspective on Studies of Productivity Growth in Asian Agriculture." In Asian Productivity Organization, *Productivity Measure-*

ment and Analysis: Asian Agriculture. Tokyo: Asian Productivity Organization.

Barker, Randolph, R. Herdt, and B. Rose. 1985. *The Rice Economy of Asia.* Washington, D.C.: Resources for the Future.

Bayly, C. A. 1983. *Rulers, Townsmen and Bazaars: North Indian Society in the Age of British Expansion, 1770–1870.* Cambridge: Cambridge University Press.

Bell, C., P. Hazell, and R. Slade. 1982. *Project Evaluation in Regional Perspective: A Study of an Irrigation Project in Northwest Malaysia.* Baltimore: Johns Hopkins University Press.

Bhaduri, Amit. 1985. "Class Relations and Commercialization in Indian Agriculture: A Study in the Post-Independence Agrarian Reforms of Uttar Pradesh." In K. N. Raj et al., eds., *Essays on the Commercialization of Indian Agriculture.* Delhi: Oxford University Press.

Bhalla, G. S. 1974. *Changing Agrarian Structure in India: A Study of the Impact of Green Revolution in Haryana.* Meerut: Meenakshi Prakashan.

Bhalla, G. S., and Y. K. Alagh. 1979. *Performance of Indian Agriculture: A Districtwise Study.* New Delhi: Sterling Publishers.

Bhalla, G. S., and G. K. Chadha. 1981. "Structural Changes in Income Distribution: A Study of the Impact of the Green Revolution in the Punjab." Report to the Centre for the Study of Regional Development, School of Social Science, Jawaharlal Nehru University, Delhi.

——. 1983. *Green Revolution and the Small Peasant: A Study of Income Distribution Among Punjab Cultivators.* New Delhi: Concept Publishers.

Bhalla, G. S., and D. S. Tyagi. 1989a. *Patterns in Indian Agricultural Development: A District Level Study.* New Delhi: Institute for Studies in Industrial Development, Indraprastha Estate.

——. 1989b. "Spatial Pattern of Agricultural Development in India." *Economic and Political Weekly* 15 (25): A46–A56.

Bhalla, Sheila. 1979. "Real Wage Rates of Agricultural Labourers in Punjab, 1961–77: A Preliminary Analysis." *Economic and Political Weekly* 14 (26): A57–A68.

Billings, Martin H., and Arjan Singh. 1970. "Mechanisation and Rural Employment." *Economic and Political Weekly* 5 (26): A61–A72.

Bliss, C. J., and N. H. Stern. 1982. *Palanpur: The Economy of an Indian Village.* Oxford: Clarendon Press.

Boisvert, Richard N., and Christine Ranney. 1990. "Accounting for the Importance of Nonfarm Income on Farm Family Income Inequality in New York." *Northeastern Journal of Agricultural and Resource Economics* 19 (1): 1–11.

Boyce, James K. 1987. *Agrarian Impasse in Bengal—Institutional Constraints to Technological Change.* New Delhi: Oxford University Press.

Brown, Lester R. 1970. *Seeds of Change: The Green Revolution and Development in the 1970s.* New York: Praeger.

Byres, T. J. 1972. "The Dialectic of India's Green Revolution." *South Asian Review* 5 (2): 99–116.

——. 1981. "The New Technology, Class Formation and Class Action in the Indian Countryside." *Journal of Peasant Studies* 8 (1): 405–454.

Cartillier, Michel. 1975. "Role of Small-scale Industries in Economic Development: Irrigation Pumpsets Industry in Coimbatore." *Economic and Political Weekly* 10 (44–45): 1732–1741.

Chadha, G. K. 1979. *Production Gains of New Agricultural Technology: A Farm Size–wise Analysis of Punjab Experience.* Chandigarh: Publication Bureau, Panjab University.

———. 1983. "Off-Farm Economic Structure of Agriculturally Growing Regions: Study of Indian Punjab." Paper presented at the conference on "Off-Farm Employment in the Development of Rural Areas," Chiangmai, Thailand, August.

Chambers, Robert, and John Harriss. 1977. "Comparing Twelve South Indian Villages: In Search of Practical Theory." In B. H. Farmer, ed., *Green Revolution? Technology and Change in Rice-growing Areas of Tamil Nadu and Sri Lanka.* Boulder, Colo.: Westview Press.

Chatterjee, B. N., et al. 1982. *How Viable Is a One Acre Farm?* Calcutta: Pearl Publishers.

Child, Frank C., and Hiromitsu Kaneda. 1975. "Links to the Green Revolution: A Study of Small-Scale Agriculturally Related Industry in the Pakistan Punjab." *Economic Development and Cultural Change* 23 (2): 249–275.

Chinnappa, Nanjamma. 1977. "Adoption of the New Technology in North Arcot District." In B. H. Farmer, ed., *Green Revolution? Technology and Change in Rice-growing Areas of Tamil Nadu and Sri Lanka.* Boulder, Colo.: Westview Press.

Chowdhury, B. K. 1970. "Disparity in Income in Context of HYV." *Economic and Political Weekly* 5 (39): A90–A96.

Clift, Charles. 1977. "Progress of Irrigation in Uttar Pradesh: East-West Differences." *Economic and Political Weekly* 12 (39): A83–A90.

Dantwala, M. L. 1973. "From Stagnation to Growth: Relative Roles of Technology, Economic Policy and Agrarian Institutions." In R. T. Shand, ed., *Technical Change in Asian Agriculture.* Canberra: Australian National University Press.

———. 1986a. *Indian Agricultural Development since Independence: A Collection of Essays.* New Delhi: Oxford and IBH Publishing.

———. 1986b. "Technology, Growth and Equity in Agriculture." In John W. Mellor and Gunvant M. Desai, eds., *Agricultural Change and Rural Poverty.* Delhi: Oxford University Press.

Dasgupta, Biplab. 1977. *Agrarian Change and the New Technology in India.* Geneva: United Nations Research Institute for Social Development.

de Janvry, Alain, and K. Subbarao. 1986. *Agricultural Price Policy and Income Distribution in India.* New Delhi: Oxford University Press.

de Soto, Hernando. 1988. *The Other Path: The Invisible Revolution in the Third World.* New York: Harper and Row.

Dev, S. Mahendra. 1988. "Prospects of Enhancing Labour Productivity in Indian Agriculture." *Economic and Political Weekly* 23 (39): A106–A113.

Divatia, V. V. 1976. "Inequalities in Asset Distribution of Rural Households." *Reserve Bank Staff Occasional Papers* (Delhi) 1 (1): 1–45.

Dreze, Jean, and Amartya Sen, eds. 1989. *Hunger and Public Action.* Oxford: Clarendon Press.

Easter, K. William, Martin E. Abel, and George Norton. 1977. "Regional Differences in Agricultural Productivity in Selected Areas of India." *American Journal of Agricultural Economics* 59 (2): 257–265.

The Economist. 1988. "An Informal Way to Grow." *Economist* 18: 95.

Etienne, Gilbert. 1982. *India's Changing Rural Scene, 1963–1979.* Delhi: Oxford University Press.

———. 1988. *Food and Poverty: India's Half Won Battle.* New Delhi: Sage Publications.

Falcon, Walter P. 1970. "The Green Revolution: Generations of Problems." *American Journal of Agricultural Economics* 52 (5): 698–710.

Finance and Development. 1989. "World Economy in Transition: The Urban Explosion." *Finance and Development* 26 (4): 47.

Food and Agriculture Organization. 1983. *Income Elasticities of Demand for Agricultural Products.* Rome: FAO.

Frankel, Francine R. 1971. *India's Green Revolution: Economic Gains and Political Costs.* Princeton: Princeton University Press.

Garg, J. S., Om Prakash, and H. L. Srivastava. 1972. "Impact of Modern Technology on Rural Unemployment." *Indian Journal of Agricultural Economics* 27 (4): 206–210.

Ghate, Prabhu. 1984. *Direct Attacks on Rural Poverty: Policy, Programmes and Implication.* New Delhi: Concept Publishing Company.

Grawe, R., J. Krishnamurti, and J. Bassh-Dwomo. 1979. *India: Employment and Employment Policy. A Background Paper.* Washington, D.C.: World Bank.

Griffin, Keith B. 1974. *The Political Economy of Agrarian Change: An Essay on the Green Revolution.* Cambridge: Harvard University Press.

———. 1989. *Alternative Strategies for Economic Development.* New York: St. Martin's Press.

Gupta, Dipankar. 1988. "Country-Town Nexus and Agrarian Mobilisation: Bharatiya Kisan Union as an Instance." *Economic and Political Weekly* 23 (51): 2688–2696.

Harriss, John. 1991. "The Green Revolution in North Arcot? Economic Trends, Household Mobility, and the Politics of an 'Awkward Class.'" In Peter B. R. Hazell and C. Ramasamy, eds., *Green Revolution Reconsidered: The Impact of the High-yielding Rice Varieties in South India.* Baltimore: Johns Hopkins University Press.

Hayami, Yugiro, and Masao Kikuchi. 1981. *Asian Village Economy at the Crossroads: An Economic Approach to Institutional Change.* Baltimore: Johns Hopkins University Press.

Herdt, Robert W. 1980. "Changing Asian Rice Technology: Impact, Distribution of Benefits and Constants." *Proceedings of the World Bank Agricultural Sector Symposia.* January: 81–127.

Hirschman, Albert O. 1958. *The Strategy of Economic Development.* New Haven: Yale University Press.

Ho, Samuel P. S. 1986. "The Asian Experience in Rural Nonagricultural Development and the Relevance for China." World Bank Staff Working Paper no. 747, Washington, D.C.

Hopper, W. David. 1965. "Allocative Efficiency in a Traditional Indian Agriculture." *Journal of Farm Economics* 47: 611–629.

India, Cabinet Secretariat. *The National Sample Survey.* No. 36, 8th round, for 1953–54; no. 144, 17th round, for 1961–62. Delhi.

India, Ministry of Agriculture. 1987. *All-India Report on Agricultural Census, 1980–81.* New Delhi.

India, Ministry of Agriculture, Directorate of Economics and Statistics. Various years. *Agricultural Situation in India.* Delhi.

——. Various years. *Agricultural Wages in India.* Delhi.

——. Various years. *Area and Production of Principal Crops in India.* Delhi.

——. Various years. *Indian Agriculture in Brief.* Delhi.

India, Ministry of Finance, Economic Division. 1989. *Economic Survey 1988–89.* Delhi.

India, Ministry of Labour, Labour Bureau. 1976. *Rural Labour Enquiry 1974–75.* New Delhi.

India, Ministry of Urban Development. 1988. *Report of the National Commission on Urbanization, August 1988.* New Delhi.

India, Office of the Registrar General. Various years. *Census of India.* General Population Tables, Series 1. Delhi.

——. Various years. *Census of India.* Village and Town Directory and Primary Census Abstract, Meerut District, Uttar Pradesh, District Census Handbook. Delhi.

India, Planning Commission. 1976. *The High Yielding Varieties Program in India: 1970–75.* Part 2. New Delhi: Programme Evaluation Organisation and Australian National University.

——. 1985. *Seventh Five-Year Plan 1985–90.* Vol. 2. New Delhi.

International Labor Office. 1974. *Sharing in Development: A Programme of Employment, Equity and Growth for the Philippines.* Geneva: ILO.

Junankar, P. N. 1975. "Green Revolution and Inequality." *Economic and Political Weekly* 10 (13): A15–A18.

Kalirajan, K. 1980. "Benefits from the High-yielding Varieties Programme and Their Distribution in an Irrigated Paddy Area." *Indian Journal of Agricultural Economics* 35 (3): 70–76.

Krishnaji, N. 1975. "Inter-regional Disparities in per Capita Production and Productivity of Foodgrain: A Preliminary Note on Trends." *Economic and Political Weekly*, special number (August): 1377–1385.

Kurian, N. J. 1987. "IRDP—How Relevant Is It?" *Economic and Political Weekly* 22 (52): A161–A176.

Ladejinsky, Wolf. 1969a. "The Green Revolution in Punjab—A Field Trip." *Economic and Political Weekly* 4 (26): A73–A82.

——. 1969b. "Green Revolution in Bihar—The Kosi Area: A Field Trip." *Economic and Political Weekly* 4 (39): A147–A167.

Lal, Deepak. 1976. "Agricultural Growth, Real Wages, and the Rural Poor in India." *Economic and Political Weekly* 11 (26): A47–A61.

Lele, Uma, and John W. Mellor. 1981. "Technological Change, Distributive Bias and Labor Transfer in a Two Sector Economy." *Oxford Economic Papers* 33 (3): 426–441.

Lenin, Vladimir I. 1960. "The Development of Capitalism in Russia: The Process of Formation of a Home Market for Large Scale Industry." In Lenin, *Collected Works*, vol. 3. 4th ed. 1899. Reprint. Moscow: Foreign Languages Publishing House.

Lerman, R., and S. Yitzhaki. 1985. "Income Inequality Effects by Income Source: A New Approach and Applications to the United States." *Review of Economics and Statistics* 67: 151–156.

Lipton, Michael, and Richard Longhurst. 1989. *New Seeds and Poor People.* Baltimore: Johns Hopkins University Press.

Lockwood, Brian, P. E. Mukherjee, and R. T. Shand. 1971. *The High Yielding Varieties Program in India—Part I.* Canberra: The Planning Commission of the Government of India and the Australian National University.

Mandal, G. C., and M. G. Ghosh. 1976. *Economics of the Green Revolution: A Study in East India.* New York: Asia Publishing House.

Marx, Karl. 1925. *Capital: A Critique of Political Economy.* Vol. 1: *The Process of Capital Production.* 1867. Reprint. Chicago: Kerr.

Meerut, Chief Development Officer. 1989. Personal communication.

Meerut, Development Authority. 1989. Personal communication.

Meerut, District Land Records Office. Collectorate.

Meerut, District Rural Development Agency. 1989. Personal communication.

Mellor, John W. 1976. *The New Economics of Growth.* Ithaca, N.Y.: Cornell University Press.

——. 1983. "Foreword." In P. Hazell and A. Roell, eds., *Rural Growth Linkages: Household Expenditure Patterns in Malaysia and Nigeria.* Research Report no. 41. Washington, D.C.: International Food Policy Research Institute.

——. 1986. "Agriculture on the Road to Industrialization." In John P. Lewis and Valeriana Kallab, eds., *Development Strategies Reconsidered.* Washington, D.C.: Overseas Development Council.

Mellor, John W., and Gunvant M. Desai. 1986. *Agricultural Change and Rural Poverty: Variations on a Theme by Dharm Narain.* New Delhi: Oxford University Press.

Mellor, John W., and Uma Lele. 1973. "Growth Linkages of the New Foodgrain Technologies." *Indian Journal of Agricultural Economics* 28 (1): 35–55.

Mencher, Joan P. 1978. *Agricultural and Social Structure in Tamil Nadu.* Bombay: Allied Publishers.

Muralidharan, M. A., et al. 1977. "An Analysis of Nutrition Levels of Farmers in Eastern Pradesh." *Indian Journal of Agricultural Economics* 32 (3): 53–60.

Nair, K. N. 1985. "White Revolution in India: Facts and Issues." *Economic and Political Weekly* 20 (25 and 26): A89–A95.

National Council of Applied Economic Research. 1965. *Techno-economic Survey of Uttar Pradesh.* New Delhi.

Neale, Walter C. 1962. *Economic Change in India: Land Tenure and Land Reform in Uttar Pradesh, 1800–1955.* New Haven: Yale University Press.

Patnaik, Utsa. 1987. *Peasant Class Differentiation: A Study in Method with Reference to Haryana.* Delhi: Oxford University Press.

Paul, Satya. 1988. "Unemployment and Underemployment in Rural India." *Economic and Political Weekly* 23 (29): 1475–1483.

Pearse, A. C. 1980. *Seeds of Plenty, Seeds of Want: Social and Economic Implications of the Green Revolution*. Oxford: Clarendon Press.

Pinstrup-Andersen, Per, and Peter B. R. Hazell. 1985. "The Impact of the Green Revolution and Prospects for the Future." *Food Reviews International* 1 (1): 1–25.

Poleman, Thomas T. 1989. "Hunger or Plenty? The Food/Population Prospect Two Centuries after Malthus." Cornell Agricultural Economics Staff Paper no. 89-30. Ithaca, N.Y.

Poleman, Thomas T., and D. K. Freebairn, eds. 1973. *Food, Population and Employment: The Impact of the Green Revolution*. New York: Praeger.

Prahladachar, M. 1983. "Income Distribution Effects of the Green Revolution in India: A Review of Empirical Evidence." *World Development* 11 (11): 927–944.

Pyatt, G., C. Chen, and J. Fei. 1980. "The Distribution of Income by Factor Components." *Quarterly Journal of Economics* 95 (3): 451–473.

Quizón, Jaime, and Hans Binswanger. 1986. "Modeling the Impact of Agricultural Growth and Government Policy on Income Distribution in India." *World Bank Economic Review* 1 (1): 103–148.

Raj, K. N., Neeladri Bhattacharya, Sumit Guha, and Sakti Padhi, eds. 1985. *Essays on the Commercialization of Indian Agriculture*. New Delhi: Oxford University Press.

Raj, K. N., Amartya Sen, and C. H. Hanumantha Rao, eds. 1988. *Dharm Narain: Studies on Indian Agriculture*. Delhi: Oxford University Press.

Raju, V. T. 1976. "Impact of New Agricultural Technology on Farm Income Distribution in West Godavari District, India." *American Journal of Agricultural Economics* 58 (2): 346–350.

Ramaseshan, S. 1988. "Management of Water Resources—Conjunctive Use: Experience in Uttar Pradesh." In J. S. Kanwar, ed., *National Seminar on Water Management—The Key to Developing Agriculture*. New Delhi: Agricole Publishing Academy.

Rao, C. H. Hanumantha. 1975. *Technological Change and Distribution of Gains in Indian Agriculture*. Delhi: Macmillan Company of India.

———. 1976. "Rapporteur's Report on Changes in the Structural Distribution of Land Ownership and Use (since Independence)." *Indian Journal of Agricultural Economics* 31 (3): 56–62.

Rao, C. H. Hanumantha, and P. Rangaswamy. 1988. "Efficiency of Investments in IRDP: A Study of Uttar Pradesh." *Economic and Political Weekly* 23 (26): A69–A76.

Rao, C. H. Hanumantha, Susanta K. Ray, and K. Subbarao. 1988. *Unstable Agriculture and Droughts: Implications for Policy*. New Delhi: Vikas Publishing House.

Reserve Bank of India. 1976. "Assets of Rural Households (as of June 30, 1971)." *In All-India Debt and Investment Survey 1971–72*. Bombay.

———. 1984. *Report of the Committee on Agricultural Productivity in Eastern India*. Vols. 1 and 2. Bombay.

Saini, G. R. 1976a. "Green Revolution and the Distribution of Farm Incomes." *Economic and Political Weekly* 11 (13): A17–A22.

——. 1976b. "Green Revolution and Disparities in Farm Incomes: A Comment." *Economic and Political Weekly* 11 (46): 1804–1806.

Saith, Ashwani, and Ajay Tankha. 1972. "Agrarian Transition and the Differentiation of the Peasantry: A Study of a West U.P. Village." *Economic and Political Weekly* 7 (14): 707–723.

Sanyal, S. R. 1988. "Trends in Landholdings and Poverty in Rural India." In T. N. Srinivasan and Pranab Bardhan, eds., *Rural Poverty in South Asia.* New York: Columbia University Press.

Sarma, J. S. 1981. *Growth and Equity: Policies and Implementation in Indian Agriculture.* Research Report no. 28. Washington, D.C.: International Food Policy Research Institute.

Schluter, M., and John W. Mellor. 1972. "New Seed Varieties and the Small Farm." *Economic and Political Weekly* 7 (13): A31–A38.

Sen, Chiranjib. 1985. "Commercialization, Class Relations and Agricultural Performance in Uttar Pradesh: A Note on Bhaduri's Hypothesis." In K. N. Raj et al., eds., *Essays on the Commercialization of Indian Agriculture.* New Delhi: Oxford University Press.

Shah, S. L., and R. C. Agrawal. 1970. "Impact of New Technology on the Levels of Income, Patterns of Income Distribution and Savings of Farmers in Central Uttar Pradesh." *Indian Journal of Agricultural Economics* 25 (3): 110–115.

Shah, S. L. and L. R. Singh. 1970. "The Impact of New Agricultural Technology on Rural Employment in North-West U.P." *Indian Journal of Agricultural Economics* 25 (3): 29–33.

Shand, R. T. 1983. "The Role of Off-Farm Employment in the Development of Rural Asia: Issues." Paper presented at the conference on "Off-Farm Employment in Development of Rural Asia," Chiangmai, Thailand, August.

Shankar, Kripa. 1978. *Uttar Pradesh in Statistics.* Allahabad: Arthik Anusandhan Kendra.

——. 1987. *Uttar Pradesh in Statistics.* New Delhi: Ashish Publishing House.

Sharma, Hari P. 1973. "The Green Revolution in India, Prelude to a Red One?" In K. Gough and H. P. Sharma, eds., *Imperialism and Revolution in South Asia.* New York: Monthly Review Press.

Sharma, Rita. 1986. "Pulses in the Food Economy of India." In J. V. Meenakshi, Rita Sharma, and Thomas T. Poleman, *The Impact of India's Grain Revolution on the Pulses and Oilseeds.* Cornell/International Agricultural Economics Study, A.E. Research 86-22. Ithaca, N.Y.: Cornell University.

Shorrocks, A. F. 1982. "Inequality Decomposition by Factor Components." *Econometrica* 50: 193–211.

Singh, Ajit Kumar. 1981. *Patterns of Regional Development—A Comparative Study.* New Delhi: Sterling Publishers.

——. 1987. *Agricultural Development and Rural Poverty.* New Delhi: Ashish Publishing House.

Singh, Baljit, and Shridhar Misra. 1964. *A Study of Land Reforms in Uttar Pradesh.* Honolulu: East-West Center Press.

Singh, Charan. 1978. *India's Economic Policy: The Gandhian Blueprint.* Delhi: Vikas Publishing House.

Singh, Daulat, et al. 1981. "Changing Patterns of Labour Absorption on Agricultural Farms in Eastern Uttar Pradesh." *Indian Journal of Agricultural Economics* 36 (4): 39–44.

Singh, Gian. 1986. *Economic Conditions of Agricultural Labourers and Marginal Farmers.* New Delhi: B. R. Publishing Corporation.

Singh, Inderjit. 1982. *Small Farmers and the Landless in South Asia.* World Bank Staff Working Paper 320. Washington, D.C.: World Bank.

——. 1988. *Land and Labor in South Asia.* World Bank Discussion Paper 33. Washington, D.C.: World Bank.

——. 1990. *The Great Ascent: The Rural Poor in South Asia.* Baltimore: Johns Hopkins University Press.

Singh, Iqbal. 1989. "Reverse Tenancy in Punjab Agriculture: Impact of Technological Change." *Economic and Political Weekly* 24 (25): A86–A92.

Singh, Katar. 1973. "The Impact of New Agricultural Technology on Farm Income Distribution in Aligarh District of Uttar Pradesh." *Indian Journal of Agricultural Economics* 28 (2): 1–11.

Srivastava, Uma K., Robert W. Crown, and Earl O. Heady. 1971. "Green Revolution and Farm Income Distribution." *Economic and Political Weekly* 6 (52): A163–A172.

Stark, O., J. Taylor, and S. Yitzhaki. 1986. *Remittance and Inequality.* Discussion Paper no. 1212. Harvard Institute of Economic Research, Harvard University.

Staub, William J., and M. G. Blase. 1974. "Induced Technological Change in Developing Agricultures: Implications for Income Distribution and Agricultural Development." *Journal of Developing Areas* 8 (4): 581–596.

Stokes, Eric. 1978. *The Peasant and the Raj: Studies in Agrarian Society and Peasant Rebellion in Colonial India.* Cambridge: Cambridge University Press.

Stone, Ian. 1984. *Canal Irrigation in British India.* Cambridge: Cambridge University Press.

Stuart, Alan. 1954. "The Correlation Between Variate-Values and Ranks in Samples from a Continuous Distribution." *British Journal of Statistical Psychology* 7: 37–44.

Subbarao, K. 1980. "Institutional Credit, Uncertainty and Adoption of HYV Technology: A Comparison of East U.P. with West U.P." *Indian Journal of Agricultural Economics* 35 (1): 69–90.

Sukhatme, P. V. 1977. "Malnutrition and Poverty." Ninth Lal Bahadur Shastri Memorial Lecture, Indian Agriculture Research Institute, New Delhi.

Sundaram, K. 1984. "Registrar General's Population Projections 1981–2001: An Appraisal and an Alternative Scenario." *Economic and Political Weekly* 19 (34): 1479–1484.

Tewari, R. N. 1970. *Agricultural Development and Population Growth: An Analysis of Regional Trends in U.P.* Delhi: S. Chand and Sons.

Todaro, Michael P. 1973. "Industrialization and Unemployment in Developing Nations." In T. T. Poleman and D. K. Freebairn, eds., *Food, Population and Employment: The Impact of the Green Revolution.* New York: Praeger.

Tyagi, S. S. 1988. "Walidpur: Agricultural Transformation in Two Decades." Research Study 88/3, Agro Economic Research Centre, Delhi University.

Uttar Pradesh, Board of Revenue. Various years. *Agricultural Census of Uttar Pradesh*. Lucknow.

——. Various years. *Crop and Season Reports of Uttar Pradesh*. Allahabad: Government Press.

Uttar Pradesh, Department of Agriculture, Directorate of Agricultural Statistics. Various years. *Agricultural Statistics of U.P., 1961–62 to 1983–84*. Lucknow.

Uttar Pradesh, Department of Agriculture, Directorate of Agricultural Statistics and Crop Insurance. Various years. *Uttar Pradesh Ke Krishi Ankre* [Agricultural Statistics of Uttar Pradesh]. Lucknow.

Uttar Pradesh, Department of District Gazetteers. 1965. *Meerut*. Allahabad: Government Press.

Uttar Pradesh, Institute of Agricultural Sciences. 1974. "Cost and Return of Important Crops in U.P." Bulletin no. 1 (April). Kanpur.

Uttar Pradesh, State Planning Institute, Department of Planning. 1988. *Planning Atlas of Uttar Pradesh*. Lucknow.

Uttar Pradesh, State Planning Institute, Economics and Statistics Division. Various years. *Sankhyiki Patrika, Meerut*. Lucknow.

——. Various years. *Statistical Abstract of Uttar Pradesh*. Lucknow.

——. Various years. *Statistical Diary of Uttar Pradesh*. Lucknow.

Vaish, R. R. 1964. *Walidpur*. Agro Economic Research Centre Report 54. Delhi: Delhi University.

Vyas, V. S. 1979. "Some Aspects of Structural Change in Indian Agriculture." *Indian Journal of Agricultural Economics* 34 (1): 1–18.

Wade, R. 1980. "India's Changing Strategy of Irrigation Development." In E. W. Coward, Jr., ed., *Irrigation and Agricultural Development in Asia*. Ithaca, N.Y.: Cornell University Press.

Westley, John R. 1986. *Agriculture and Equitable Growth*. Boulder, Colo.: Westview Press.

Whitcombe, Elizabeth. 1972. *Agrarian Conditions in Northern India*. Berkeley: University of California Press.

Wiser, William H., and Charlotte Viall Wiser. 1963. *Behind Mud Walls, 1930–1960*. Berkeley and Los Angeles: University of California Press.

World Bank. Various years. *World Development Report*. Washington, D.C.: World Bank.

Yamada, Saburo. 1987. "Agricultural Growth and Productivity in Selected Asian Countries." In Asian Productivity Organization, *Productivity Measurement and Analysis: Asian Agriculture*. Tokyo.

Index

Agro-Economic Research Center (AERC), 63, 69–70

credit
  dudhiya as source. *See dudhiya* (milk vendor)
  governmental encouragement, 249–250
  grain dealers as source, 146
  importance to dairying, 152, 187–188
  IRDP as source, 146–147, 151, 155, 187–188, 194–204, 228
  moneylender as source, 146
  spread of institutional, in Uttar Pradesh, 54–55

dairying
  governmental encouragement, 251–253
  importance to small producer, 181–184
  as mechanism of income diffusion in case-study villages
    factors influencing, 181, 184, 187–189
    household case studies, 194–204
    as household economic strategy, 193–194
  as mechanism of income diffusion in Meerut District, 115–117
  women's role, 178
Daurala Growth Center, 68–69
de Soto, Hernando, 245
dudhiya (milk vendor)
  in case-study villages, 152–153, 156, 166–167, 177–178

prices paid by, 184–185
  as source of credit, 184–188, 252

employment diffusion. *See* income diffusion

fertilizer
  as component of Green Revolution, 2
  trends in use in Uttar Pradesh, 49–50
Five-Year Plan
  Eighth, 245
  Seventh, 245

Gini coefficient, decomposition of, 85–89, 142–143, 191–193, 224–227, 244, 256–258
Green Revolution
  definition, 1–2
  impact
    in case-study villages, 129–133, 178–183, 213–216
    on income and employment after 1975, 18–19, 239
    on income and employment in 1965–1975, 2–17, 239
    on income diffusion. *See* income diffusion
    on Indian foodgrain production, 14–17
    in Meerut District, 106–108
    regional differences in India, 3–5
    in Uttar Pradesh, 38–58, 240–241
    in Walidpur Village, 71–89

269

high-value crops. *See also* potatoes; vegetables
    importance to rural poor, 211
    as mechanism of income diffusion in case-study villages
        factors influencing, 217–223
        household case studies, 228–233
        as household economic strategy, 226–228
    as mechanism of income diffusion in Meerut District, 117–119
high-yielding varieties. *See also specific crops*
    as component of Green Revolution, 2
    impact on smallholder viability, 16

income diffusion
    governmental encouragement, 243–254
    mechanisms. *See* dairying; high-value crops; off-farm employment
    as second-generation effect of Green Revolution, 18–19, 243
    in Walidpur Village over time, 81–86, 241
income distribution
    in case-study villages, 85–89, 140–143, 190–193, 224–227, 243–244
    impact on
        of agricultural labor, 86–89, 142–143, 192–193, 224–227
        of crop income, 86–89, 142–143, 192–193, 224–227
        of dairying, 86–89, 142–143, 192–204, 224–227
        of off-farm employment, 86–89, 142–143, 147–167, 192–193, 224–227
India
    employment problem, 12–14
    impact of Green Revolution
        on food grain production, 14–17
        on regional disparities, 3–5
    population growth, 11–14, 255
    regional differences, 2–3
Industrial Training Institutes (ITIs), 250
Integrated Rural Development Program (IRDP)
    needed changes, 249–250, 253–254
    as source of credit, 146–147, 151, 155, 187–188, 194–204, 228
    terms of lending, 146–147
irrigation
    changing importance of types
        in case-study villages, 180–181, 215–216, 221–222

    in Meerut District, 107
    in Uttar Pradesh, 33–34, 37–39
    in Walidpur Village, 76–77
    as component of Green Revolution, 2, 240
Izarpur Village
    dairying in
        factors influencing, 181–184
        household case studies, 194–204
        as household economic strategy, 193–194
    description, 64, 176–181
    impact of Green Revolution on, 178–183
    reasons for selection as case study, 120, 176

Jamalpur Village
    description, 64, 211–212
    high-value crops
        factors influencing, 217–223
        household case studies, 228–233
        as household economic strategy, 226–228
    impact of Green Revolution on, 213–216
    reasons for selection as case study, 120

land reform
    in Meerut District, 106–107
    in Uttar Pradesh, 36–37
land tenure
    in case-study villages, 72–75, 129–130, 178–179, 213–214
    in India, 6–8
    in Uttar Pradesh, 28–30

maize
    in case-study villages, 130–131, 181–182, 214–215
    production costs and returns, 217–218
mechanization
    as component of Green Revolution, 8, 108
    impact of
        on agricultural wages, 8–9, 82
        on rural employment, 8–9, 82
        on women, 183
Meerut District
    applicability of development experience, 242–243
    compared with other areas in Uttar Pradesh, 108–110
    description, 64, 105–108

Meerut District (*cont.*)
  impact of Green Revolution on, 106–
    108
  mechanisms of income diffusion, 110–
    119
milk. *See also* dairying
  marketing, 184–187
  nutritional benefits, 117

off-farm employment
  governmental encouragement, 244–251
  as mechanism of income diffusion in
    case-study villages
    factors influencing, 136–138
    as household economic strategy, 131–
      140, 144–147
    types, 133–134, 138–140
  as mechanism of income diffusion in
    Meerut District, 110–114
  rural growth centers as locus, 137–138
oilseeds, adverse impact of Green Revolution on
  in case-study villages, 77–79, 130–131,
    181–182, 214–215
  in Uttar Pradesh, 44

population
  in case-study villages, 129, 178, 212
  in India, 11–14, 255
  in Meerut District, 106
population pressure on land
  in India, 6–14
  in Uttar Pradesh, 28–30
potatoes
  cold storage, 222
  demand elasticities, 115, 217
  labor requirements, 217–218
  marketing problems, 222
  production costs and returns, 217–218
  production in Uttar Pradesh
    in case-study villages, 77–79, 181–
      183, 214–233
    East-West changes in production, 43–
      44
    in Meerut District, 117–119
poverty line
  defined, 84
  impact of Green Revolution on, 9
  population with incomes below
    in case-study villages, 84, 142, 190–
      191, 224, 243
    in India, 9
pulses
  adverse impact of Green Revolution on

    in case-study villages, 77–79, 130–
      131, 181–182
    in Uttar Pradesh, 44
  production costs and returns, 217–218

Rampur Village
  description, 64, 127–131
  impact of Green Revolution on, 129–
    133
  off-farm employment
    factors influencing, 136–138
    household case studies, 147–154,
      157–163
    as household economic strategy, 131–
      140, 144–147
    types, 133–134, 138, 140
  reasons for selection as case study, 120
rice (paddy) production
  in case-study villages, 77–79, 130–131,
    214–215
  in India, 15
  in Uttar Pradesh
    changed competitive position, 43–44
    East-West changes in production, 43–
      44
    yield changes, 44–45
roads and transportation
  governmental encouragement, 245–246
  in Uttar Pradesh, 35, 55
rural electrification
  problems of reliability, 108, 161–164,
    246–247
  in Uttar Pradesh, 54
  in Walidpur Village, 66
rural growth centers
  changes in importance in Uttar Pradesh,
    50–54
  Daurala Growth Center, 66–69
  governmental encouragement, 247–249
  influence on off-farm employment in
    case-study villages, 137–138, 177
  in Meerut District, 113–114
  as source of off-farm employment, 113–
    114
  in Uttar Pradesh, 50–54
rural poverty. *See* poverty line

sharecropping. *See* tenancy
Sitapur Village
  household case studies of off-farm em-
    ployment, 154–157, 163–167
  reasons for selection as case study, 147–
    148

sugarcane
    appeal to large farmers, 76, 78
    in case-study villages, 68, 77–79, 130–
        131, 181–182, 214–215
    in Meerut District, 107–109
    production costs and returns, 217–218

tenancy, impact of Green Revolution on,
    in case-study villages, 73–75, 226–
    228
Training Rural Youth for Self-Employment
    (TRYSEM), 147, 199, 251
tubewells
    as component of Green Revolution, 2,
        240
    cost of in Jamalpur Village, 221
    growing importance
        in case-study villages, 180–181, 215–
        216
        in Meerut District, 107
        in Uttar Pradesh, 37–39, 240. *See also*
        irrigation: in Uttar Pradesh
    in Walidpur Village, 76–77

Uttar Pradesh
    applicability of development experience,
        242–243
    description, 23–30
    East-West differences, 30–38
    impact of Green Revolution
        period of narrowing disparities, 38–
        58, 240–241
        period of widening disparities, 38–58,
        240–241
    population growth, 23, 50–54
    rural growth centers, 50–54

vegetables
    demand elasticities for, 115, 217
    increases in production
        in case-study villages, 77–79, 130–
        131, 181–183, 217, 233
        in Meerut District, 117–119
        in Uttar Pradesh, 44
    production costs and returns, 217–218

Walidpur Village
    compared with Uttar Pradesh and
        Meerut District, 67–68
    description, 63–68
    impact of Green Revolution
        on agriculture and land use, 71–81
        on household income, 81–89
        on poverty, 84, 243
    socioeconomic surveys
        benchmark survey of 1963–1964, 69–
        70
        survey of 1988–1989, 70–71
wheat production
    in case-study villages, 77–79, 130–131,
        181–182, 214–216
    costs and returns of, 217–218
    in India, 15
    in Meerut District, 107
    in Uttar Pradesh
        changed competitive position, 42–43
        East-West changes in production, 42–
        43
        yield changes, 44–45
women
    changed role as result of mechanization,
        183
    role in dairy production, 178, 188–189
    as sources of credit, 187

Food Systems and Agrarian Change

Edited by Frederick H. Buttel, Billie R. DeWalt,
and Per Pinstrup-Andersen

Hungry Dreams: The Failure of Food Policy in Revolutionary Nicaragua, 1979–1990
by Brizio N. Biondi-Morra

Research and Productivity in Asian Agriculture
by Robert E. Evenson and Carl E. Pray

The Politics of Food in Mexico: State Power and Social Mobilization
by Jonathan Fox

Searching for Rural Development: Labor Migration and Employment in Rural Mexico
by Merilee S. Grindle

Structural Change and Small-Farm Agriculture in Northwest Portugal
by Eric Monke et al.

Diversity, Farmer Knowledge, and Sustainability
edited by Joyce Lewinger Moock and Robert E. Rhoades

Networking in International Agricultural Research
by Donald L. Plucknett, Nigel J. H. Smith, and Selcuk Ozgediz

The New Economics of India's Green Revolution: Income and Employment Diffusion in Uttar Pradesh
by Rita Sharma and Thomas T. Poleman

Agriculture and the State: Growth, Employment, and Poverty in Developing Countries
edited by C. Peter Timmer

Transforming Agriculture in Taiwan: The Experience of the Joint Commission on Rural Reconstruction
by Joseph A. Yager

Library of Congress Cataloging-in-Publication Data

Sharma, Rita, 1950–
    The new economics of India's Green Revolution : income and
employment diffusion in Uttar Pradesh / Rita Sharma and Thomas T.
Poleman.
      p.   cm. — (Food systems and agrarian change)
    Includes bibliographical references and index.
    ISBN 0-8014-2806-8
    1. Green Revolution—India—Uttar Pradesh. 2. Agriculture—
Economic aspects—India—Uttar Pradesh. 3. Agriculture and state—
India—Uttar Pradesh. I. Poleman, Thomas T. II. Title.
III. Series.
HD2075.U6S49   1993
338.1'0954'2—dc20   92-56784